BREAKING THE CODE

BREAKING THE CODE

FIVE STEPS TO A LIFE-CHANGING
SOFTWARE DEVELOPMENT JOB

BOBBY DAVIS JR.

LIONCREST
PUBLISHING

COPYRIGHT © 2020 BOBBY DAVIS JR.
All rights reserved.

BREAKING THE CODE
Five Steps to a Life-Changing Software Development Job

ISBN   978-1-5445-0952-5 *Hardcover*

978-1-5445-0951-8 *Paperback*

978-1-5445-0950-1 *Ebook*

978-1-5445-0953-2 *Audiobook*

To Kelly, the love of my life. You have followed me down every crazy path I've taken and supported me on every adventure I have wanted to embark on. I look forward to many more adventures with you.

CONTENTS

Disclaimer: Some names and identifying details have been changed to protect the privacy of individuals.

INTRODUCTION

Three years ago, a teacher at Western Alamance Middle School invited me to speak to her class on career day. She wanted me to introduce her students to the software development industry. Before my turn to speak, representatives from many other fields spoke positively about their work. They were, to a person, successful by most accounts, but none of them could deliver the speech I was about to give.

The last speaker sat down, and I stood up in front of those young, aspiring students to tell them the truth I knew they needed to hear.

"I'm sure all of the jobs you've heard about today can lead to success. There are, after all, many ways you can achieve your career ambitions. However, only one job, software development, can maximize your opportunities in the modern economy.

"Once you learn to code, software development can be your ticket to any life you want to live. With a coding job, you can make as much as you want and live where you want. You can work for any company on the planet. If you want, you can even take over the world."

Silence fell over the room as that idea sunk in. Then, I gave them a real shock.

"I can tell you with supreme confidence that the next Mark Zuckerberg is sitting in a classroom exactly like this one right this minute. All it takes to achieve Zuckerberg's kind of success is a good idea and the ability to code. With that, you can become the next tech billionaire."

Some of the class laughed out loud in disbelief. For them, Facebook had always been a tech giant, and Zuckerberg forever lauded as a visionary. They felt that they were as likely to be the next Mark Zuckerberg as the next Harry Potter. Even the teacher couldn't believe it.

Most of us forget that Facebook's origins are far from grand. As recently as 2004, Zuckerberg was just another 20-year-old undergrad, sitting in his dorm room, wishing he was out enjoying the parties around campus. However, instead of spending his evenings watching TV like most bored college kids, he developed the first iteration of a social networking platform he called FaceMash. His aim, at that point, wasn't to take over the world or change how we communicate; it was to meet more girls.

The world's most popular social media site developed out of that lonely Harvard student's desire to become more popular. With that simple aspiration and some decent coding skills, Zuckerberg went from a nobody to one of the wealthiest and most influential people in the world.

Facebook wasn't an ingenious insight into society or the human condition. College students have been looking for novel ways to find dates for decades. I'm sure plenty of Zuckerberg's own classmates were sitting around that same semester, wondering where the party was. The only difference between them and Zuckerberg was that he could code, and they couldn't.

The rest is history.

Summing up my career day speech, I told those middle school kids, "The economy of today runs on software. When you learn how to code, you open the door to every opportunity because every business needs your skills. Whether you want to be Zuckerberg or just have the best job on your street, it all starts with learning how to code and how to get into this industry.

"The American dream still exists; it's just written in code."

THE SECRET TO MODERN SUCCESS

The world of software development is extensive enough to accommodate everyone's dreams. The next world-changing tech entrepreneur is already at work using coding skills to realize their ideas and sell them to the world. At the same time, there are thousands upon thousands of well-established companies across this country looking to make their new software developer the best-paid employee on their payroll. Every industry in every town in America needs coders to create apps and websites for them. And they need those positions filled urgently. Once you have the skills and know-how to navigate the software development industry, you can have your pick of these positions.

Moreover, success in this field doesn't require genius-level skills and insights. Because the modern world runs on software, you no longer have to be a Bill Gates or Steve Jobs to become an overnight tech millionaire or to find a steady $120,000-a-year job with a 401(k).

With a good work ethic and decent coding skills, you can work in the job you want, for the company you want, at the wage you want, in the place you want.

A NEW–AND BETTER–ECONOMY

Over the last two decades, the job market as we used to know it has begun to disappear. The old jobs we thought we could rely on, in industries such as manufacturing, mining, and construction, are gone. Thanks to automation, other big employers in retail, transportation, and customer service may soon follow. It's a new economy, and many of us are feeling left behind.

You don't need to fear this new job market; it's full of opportunity. However, this opportunity now lies in a different industry. And, believe it or not, that's a good thing.

The old jobs in manufacturing and service didn't care about your creativity or intellect. They paid you a decent (though hardly spectacular) wage for using your physical abilities. The brainy person who fixed factory equipment in previous generations wasn't the most highly regarded or the highest paid employee on the floor. The IT department that kept the phones working at call centers twenty years ago was just as underappreciated as the people on the phones.

Nowadays, that brainy person runs the entire software development shop for the company, and they are rewarded handsomely for it. Learning how to code opens up a world that you may not have known even existed, a world in which you are the most in-demand employee in town. You no longer have to be trapped in any job or location; you no longer have to settle for unfulfilling work where no one appreciates you.

At Coder Foundry—our boot camp that specializes in teaching novices how to code—we constantly and consistently place new software developers in lucrative, fast-advancing positions. Every month, we introduce individuals to the potential that has been waiting for them.

Many of our students are career-switchers—people who have spent years underemployed, underpaid, and under-fulfilled in jobs they had once thought of as "safe" or "steady." Among these same students are naturally intro-verted individuals that society has labeled "slackers" or "lazy." That label is wrong. These individuals aren't lazy; they just have personality types more suited to working independently, unlike extroverts. Before finding out about software development, they didn't see a way into profit-able positions that would suit their natural interests, so they never tried.

The underappreciated career-switchers and the misla-beled slackers come in all ages and from all backgrounds. At Coder Foundry, we welcome young people looking for a way to get their careers started, as well as middle-aged people who have seen their careers halted. Our classes are full of people from rural America, the inner city, and the suburbs.

These individuals are, to a person, smart problem-solvers with the motivation to succeed. They simply couldn't find a career that offered them a way to make the most of their talents. With our help, they've discovered that the problem wasn't them; it was the work they were pursuing.

YOU CAN HAVE A GREAT TECH JOB, TOO

One thing everyone mistakenly seems to agree on about the modern economy is that tech jobs are great, but they're impossible to get. Everyone dreams of working in one of those high-paying, influential, innovative positions in Silicon Valley. The jobs are cool, rewarding, and even a gateway to power, but they're not for people like you and me. This assumption is all wrong.

The only thing that sets those Silicon Valley millionaires and billionaires apart from the rest of us is their ability to code. Their coding skills gave them a leg up. It's what made each of them a success in the new economy. Without those skills, they would be just like the rest of us: struggling for a way into the American dream.

But guess what? You can learn how to code, too. Coding isn't exclusively for Ivy League students or young people with big-city connections. Software development isn't like Hollywood or Wall Street. It isn't about where you went to school or who you know.

With a little guidance, you too can enter this lucrative world. You too can go home with a big paycheck and a world of advancement potential. And, if you want, you too can get a position in Silicon Valley. Or, you can find an equally great job right next door. It's all up to you.

FROM $500 IN THE BANK TO SUCCESS

A lot of people may look at the biographies of tech billionaires and say, "Sure, coding made them a success, but they're special. I'm not like them."

The reality, though, is that there's very little special about Mark Zuckerberg or his biography. He wasn't a once-in-a-generation genius like Bill Gates. He wasn't a visionary like Steve Jobs. His business decisions and policies aren't even particularly admirable. He just had a basic idea, and he had the knowledge required to implement it.

And he's not the only one to do it, either. Thousands of tech entrepreneurs you've never heard of have done the same thing on only a slightly smaller scale. I've done it myself.

In 2002, I was sitting in a Barnes & Noble drinking coffee with my future business partner, Dave DeBald. Both of us had just been let go by our former employer, who had downsized their tech department at the height of a recession. With just $500 in the bank, I could barely justify the price of the coffee. Dave felt like we were commiserating over the end of our software development careers.

But I wasn't ready to give up. I told him, "Let's start our own software development company."

"We can't just start a company," he said. "Can we?"

"Why not? How hard can it be?"

The answer was not hard at all. We transferred that last $500—minus the coffee—into a business account, came up with the name Core Techs for our company, and called up all our former employer's clients.

"Remember us?" we asked. "We're now working for Core Techs. Would you like to switch your invoicing over to us?"

Most of them did, and just like that, we had a successful business. Within two days, we were fully employed again. Even at the height of a recession, people needed our coding skills. Dave and I have been working for Core Techs ever since, which has gone on to become a multimillion-dollar company.

We created another successful business in 2007 when we founded Advanced Fraud Solutions. That's probably another company you've never heard of, but it now makes $10 million a year by protecting financial institutions in forty-eight states.

Like Zuckerberg, Dave and I aren't special. We just had

the right skills for the current job market. I can walk into any café in the country today and find dreamers who are ready to create their own companies. I can walk into any office in the country and find problem-solvers who could rise to the highest levels of the biggest corporations on the planet. I can walk down almost any street and find doers who could be making $10,000 a month keeping local businesses competitive.

The only reason these dreamers, problem-solvers, and doers aren't rising up the economic ladder is because they don't know how to code or how to take the knowledge of coding and get into our industry.

At least, they don't know yet.

FOUNDING A PATH TO CODING SUCCESS

Those dreamers, problem-solvers, and doers are the reason I founded Coder Foundry. Our mission has always been to gather together the bright, talented, underutilized individuals struggling in the job market and place them in positions where they can finally thrive. We've been astoundingly successful at that. Eighty-five percent of our students graduate into lucrative jobs full of future potential.

I'd love to claim that I am the reason so many of my students have been successful, but it isn't really about me. I didn't change these people's lives. I simply showed them the path forward. They learned to walk the path themselves.

If anything, my main role at Coder Foundry has been to demystify the software development industry. There is a clear path that leads from learning how to code to landing your first job and setting yourself up for unlimited success. Most people just don't know that path exists.

That's what this book is going to teach you. I'll walk you through the five steps you have to take to get from where you are today—with a job that undervalues and underpays you—to your first well-paying job. Then, I'll show you how you take that initial position and develop your career to reach the job of your dreams—in any town, any industry, earning as much as you want.

You'll learn a lot in this book, but one thing you won't learn is how to code. This book isn't a coding manual. It won't teach you coding languages or techniques. For that, you'll need Coder Foundry or another boot camp.

Instead, I'm going to teach you how to become the next Zuckerberg. Or, if you prefer, how to gain a stable, fulfilling job that allows you to become the kind of person you always wanted to be: driving back from a job you love to a home in an area of the country you love. The choice is yours, but it will be your choice, so long as you follow the advice ahead.

You don't have to be a genius. You don't have to go to Harvard. You just need to follow these steps and stick to your new craft. That's all it takes for a life of great opportunity to open up to you.

I've seen it happen hundreds of times. It can happen to you, too.

PART I

WHY CODING?

———

CHAPTER 1

THE ECONOMIC MOBILIZER OF OUR TIME

Five years ago, Richard Newton was a twenty-six-year-old living in his parents' basement. Like millions of others who have struggled longer than most to start a career, he felt bad about himself. Some would have referred to Richard by the unfortunately common derogatory term "loser." I would have referred to him by the far more accurate term "failure to launch." The distinction between these terms is important. Richard wasn't destined to be a failure. He wasn't lazy, ignorant, or incurious. He simply hadn't yet found his path to success.

Then he heard about Coder Foundry, and he decided to check it out. During our first conversation, he told me that he didn't believe he would ever make much of his life.

"I'm an introvert," he said. "People don't hire me. They make fun of me."

Over the years, people had told him there was something wrong with him. They had said success was for other people. He had internalized this negative view of himself, and so he'd stopped trying to move forward.

Luckily, something about the software development sphere had attracted his attention, and despite his reservations about himself, he became one of my first students.

Within weeks of finishing his course, he proved the naysayers wrong. He landed a job at a local garage door company for $51,000 a year and immediately started making more annually than most of the people who had made fun of him. After three months in our boot camp, he'd turned his whole life around.

"I didn't believe anybody would ever hire me," he told me, with tears in his eyes.

I responded to him with the truth: "Of course they hired you. You're exactly what they've been looking for."

Richard's story is not unique. At Coder Foundry, we've helped start the careers of hundreds of "failure to launch" individuals.

The truth is that never before in the history of the world has there been a career path that offers so much life-changing potential to so many individuals. No matter where you are and no matter your background or personal history, if you are clever, creative, and dedicated, software development can provide you with economic opportunities beyond your wildest dreams—just as it did for Richard.

THE SERVICE ECONOMY NEEDS SOFTWARE DEVELOPERS

The rise of the software developer has coincided with a huge shift in the American economy.

America used to be a manufacturing economy. For more than a century, the factory was the center of American economic life. As recently as 1970, more than one in every four working individuals was employed by the country's factories.[1] Now, it's one in ten.

From the mid-nineteenth century to the mid-twentieth century, the factory represented the chance for a better life for individuals and their families. People abandoned their farms and fields for the city and the factory, where they could make far more sweating over a machine than sweating over the land. The factory represented the chance

---

1   "Percent of Employment in Manufacturing in the United States (DISCONTINUED)," Federal Reserve Bank of St. Louis, June 10, 2013, https://fred.stlouisfed.org/series/USAPEFANA.

to escape poverty, enter the middle class, buy a house, raise kids with a better education, retire at a reasonable age, and live on a pension. It's not hard to imagine why the loss of those jobs has affected people so deeply.

Unfortunately, the nature of those jobs has changed, and most likely, they aren't coming back. In fact, they can't come back because they haven't gone anywhere. The work still exists, it's just done by robots now. Ford hasn't stopped making cars in America; they've simply programmed robots to replace many of the factory workers.

Since the 1950s, America has been transitioning away from a manufacturing economy to a service economy. Today, 70 percent of Americans are involved in the service industry in one aspect or another, outnumbering factory workers seven to one.[2]

---

2   Drew DeSilver, "10 Facts about American Workers," *Fact Tank News in Numbers*, Pew Research Center, August 29, 2019, https://www.pewresearch.org/fact-tank/2019/08/29/facts-about-american-workers/.

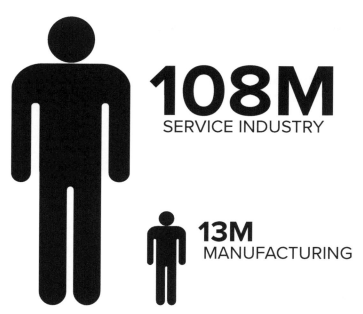

108M
SERVICE INDUSTRY

13M
MANUFACTURING

The service industry now dominates manufacturing.

At this point, the service economy looks to be here to stay, which is actually great news for you. The service economy offers far greater potential rewards than the factories ever did. You just need the right skills to reach those rewards.

As service enters a new era and transitions into the digital environment, every business has to connect to customers through websites, apps, and digital communication. To do that, they need software developers. All the services customers expect from businesses online—from checking accounts and paying bills to tracking when their pizza will be delivered—require software developers.

Over the past two decades, we've transitioned into a full-fledged, app-based, online service economy, and every technology-based interaction between a business and a customer (or a business and another business, for that matter) requires software developers to facilitate those interactions.

In this way, software development has become the new factory job. It is widely available, in high demand, and offers a leap up the social ladder. This is the fastest growing industry in America, and companies from Honolulu to San Juan, Puerto Rico, and from Juneau to Key Largo are desperate to fill these positions in their offices. They're just as desperate in little places like my hometown of Kernersville, North Carolina, as they are in New York City.

THE THREE ECONOMIC MOBILIZERS

Software development may be the new factory job, in terms of its ubiquity and demand, but the benefits that come with a coding job far exceed anything the factory could ever offer. To work in a factory, you have to live in one place, work for one employer, and settle for the wages the union can bargain for you.

If you lived in Michigan fifty years ago, your best option was to go to Detroit. There, you could work in one of the car factories. You'd join the union and get the same wages as everybody else on the floor. And that was it. If you'd rather live in Florida, it didn't matter. The job was in Detroit. If you'd rather work in a steel factory, you could forget that dream as well. Steel was over in Pittsburgh, and there was little guarantee you could make enough to afford the move. At best, you could hope to move up into floor management and boost your salary a bit over the coming decades.

Software development offers so much more on every front. With a coding job, you have complete economic mobility. You aren't tied to one place, one company, or one salary. You can make as much as you want, with whatever company you want, and where you want.

FINANCIAL MOBILITY

How often have you dreamed of making $100,000 a year? How quickly do you tell yourself to wake up and stop dreaming about fairy tales?

If you became a software developer, you wouldn't have to dream about a six-figure income. You'd be earning it. The median income for a software developer is $103,620 a year. That's more than double the average wage in America. Software developers can command that kind of income because their job has been the most in-demand job in the country for three years running. At this moment, there are almost a quarter of a million software development jobs in the country right now. The unemployment rate in this area is 1.6 percent.[3]

Think of all the things you have wanted to do with your life that you couldn't because of financial limitations: buy a house, travel the world, drive an expensive car, volunteer for a cause you believe in. Everything is within your financial reach in this career. You'll join the list of people who live life on their terms because money is no longer a constraint; it's an enabler.

JOB MOBILITY

Many jobs that offer financial mobility come at a price: you have to take undesirable positions and stick with them. They require long hours, stressful work conditions, and little personal time. There's little room for creativity or personal fulfillment. That's not the case with software development.

If you are in a software development job you don't like, you can simply leave and find another position. You'll be in such high demand, you can choose your employer, not the other way around.

So, if your company has a toxic culture or you don't like your boss, all you have to do is leave. You'll have another job by

---

3   "Software Developer Overview," *Best Jobs Rankings*, *U.S. News & World Report*, https://money.usnews.com/careers/best-jobs/software-developer.

the time you get home, one that better fits your professional expectations. If you don't like the hours or the stress level at your current job, it's easy to find a more laid-back employer who will pay you the same salary.

Not only are you not tied to a specific company or work environment, you aren't tied to a specific industry. Do you want to work on Wall Street? They need software developers to incorporate machine learning into predicting stock futures. Do you want to work for a nonprofit trying to save the environment? They need apps that can track endangered species. Since every business needs software developers, there are no limitations on where you can employ these skills.

GEOGRAPHIC MOBILITY

Everyone has a place they've always wanted to live. For some, it's bustling New York City. For others, it's easygoing Southern California. For me, it's Kernersville, North Carolina. Wherever your dream home is, with software development, you can live there.

Software development jobs exist in small towns in the Midwest and downtown metropolises along every coast. And if for some reason the place you want to live doesn't have a position, software developers can work remotely.

You can work from home or in an office. You can work while

you travel the world. So long as you've got a steady internet connection, there are no limitations on your location.

A CAREER DESIGNED FOR YOU

Software development can offer you stability, mobility, and some of the best wages for any career in America. With this job, you can have it all. And I do mean *you*.

One of the main excuses people give me for not signing up with Coder Foundry is that companies won't hire someone like them. Because of their skin color, gender, or lifestyle, they believe no amount of coding skill will allow them to enjoy these benefits.

Those assumptions deny an untold number of potential software developers the chance at a better life, and the worst part about those assumptions is that they are *not* true. The graduates we place in positions across America are extremely diverse. This is a skills-based industry, and the vast majority of businesses are interested solely in how well you can perform those skills. Remember, these businesses are desperate for your services. If they don't hire you, they'll look to India, Nigeria, and the Philippines to fill that position. Bringing software developers to America represents a great expense for their business, so if they can hire you locally, they will put aside any other concerns to do so.

The same is true for those who feel they are too old for this work. In software development, there simply is no such thing as "too old." According to the American Economic Review, the average age of a tech startup founder is not twenty or even thirty, it's forty-three.[4] That's just an average. Plenty of those now-successful entrepreneurs are in their fifties and sixties. Some must be even older. Just because the impression the media gives you is that every tech success story involves a brilliant, young individual camped out in Silicon Valley doesn't mean that's the truth. In software development, there is plenty of room for people of all ages, backgrounds, sexes, races, and personality types.

And if somehow no one will hire you because of who you are, this career gives you the opportunity to hire yourself and prove them all wrong. You can design the next great online service, app, or game on your own, and you can reap all those benefits yourself.

No matter your age, background, or beliefs, if you are a hardworking, creative problem-solver, this is a job designed for you. The economic mobilizers are there for you. All you need now are the skills and the know-how to navigate the industry.

---

4   Pierre Azoulay et al., "Age and High-Growth Entrepreneurship," *American Economic Association: Insights*, https://www.aeaweb.org/articles?id=10.1257/aeri.20180582&&from=f.

CODING CAN BE YOUR NEW BEGINNING

When fifty-two-year-old Tom Harrison came to Coder Foundry, he had real anxiety that it was too late for him to start over in a new field. He had spent his career working as a financial planner in a bank, but after decades behind the desk, he still wasn't making a good salary. Tom wanted a change, but he worried it was no longer possible. He asked me directly, "Can I really do this at my age?"

The answer, as with Richard before him, was a resounding yes. Since completing his course, Tom has been working consistently and successfully in software development. He's now designing the financial services software that he would have used as a financial planner, except now, he's making the salary he always dreamed of.

Before coming to Coder Foundry, both Tom and Richard felt life and success were going to pass them by. They felt that big salaries and fulfilling positions would always go to someone who was more social or younger. They were both wrong.

There's room in this industry for the brilliant and the competent, the entrepreneur and the steady nine-to-fiver, the young and the old, and people of every background.

There's plenty of room for you, too. The potential careers I've sketched out for you in this chapter may sound like a dream, but this is a reality thousands of software developers are already living. To place you on the path to joining them, let's clear up the mystery behind the technology and get down to the facts about what coding is and why it's the most in-demand skill in the world.

{ CHAPTER 2 }

WHAT IS CODING?

When I spoke to the students at Western Alamance Middle School, I didn't expect all of them to grow up to be the next Mark Zuckerberg. We don't need a million Facebooks in the world. However, we do need software developers at every level of the economy. There are plenty of amazing, interesting software development positions for everyone in that classroom and in classrooms and living rooms across the country.

Walk in the doors of any successful organization today, and you will find coders at work. Coder Foundry alone has placed hundreds of former students in every industry imaginable.

One of our graduates, Devon Smith, went from being a local bartender to helping Duke Energy manage the electrical grid along the Atlantic coast.

Another student, Jared Thomas, was hired by a large pharmaceutical company in Brooklyn, New York. He is now the development manager for their entire coding department.

Sandra O'Keefe was yet former another student of ours. Before Coder Foundry, she worked as a hairdresser. Now, she's part of the coding team at a large multistate car dealership with thirty-eight locations.

This success isn't limited to Coder Foundry graduates. One of my friends, Michael Turner, was an avid soap maker. He and his wife loved to come up with new homemade soap recipes. Because Michael was a coder, he was able to turn that hobby into the website soapcal.net, where he provided recipes, advice, and a measurement guide to his fellow amateur soap makers. Before he sold the site for a fortune last year, soapcal.net was making his family $12,000 a month.

This is a job that can take you anywhere. You can work from home like Michael or in New York City like Jared. You can work for big Silicon Valley companies like Apple or a Redmond, Washington, company like Microsoft. You can code for gaming companies or electrical companies. You can work on the latest AI with Tesla or the U.S. military.

Whatever direction you wish your career to go, coding makes it possible. But in order to take advantage of any of these opportunities, you have to first understand what coding is and how it became the key to success in our economy.

THE MEANING OF CODE

Before you can head down the path to a successful career in software development, you'll need to understand the basic terminology within the industry. And that starts by understanding what coding actually means.

Code is, at the most basic level, a language. More precisely, it is a huge set of languages that are intelligible to computers. Coding, then, is the act of "speaking" to a computer by sending it instructions to perform an action.

```
public ActionResult Create(int? projectId)
{
TicketCreateViewModel viewModel = new TicketCreateViewModel();
Ticket ticket = new Ticket();
if (projectId != null)
{
    viewModel.Project = db.Projects.Find(projectId);
}
viewModel.User = db.Users.Find(User.Identity.GetUserId());
viewModel.Projects = projectHelper.ListUserProjects(viewModel.User.Id);
viewModel.Ticket = ticket;
return View(viewModel);
}
```

A language (in this case, C#) that a computer can understand.

To make a computer perform any action, you have to "speak" to it in a language it understands. To you and me, these languages can look almost alien, but to a computer, they represent clear commands.

This is really no different than how we respond to language in our lives. If someone asked me in English to stand up and walk across the room, I'd be happy to oblige them. But if they made the same request in Arabic or Swedish, I wouldn't move because I wouldn't understand the request. Your computer works the same way. You can't make a computer open a program or run a calculation by speaking to it in English—at least not directly. The computer doesn't speak English; it speaks hundreds of languages of code.

Whether you're writing an essay in Microsoft Word, sending an email in Gmail, or playing a game on your iPhone, someone has created those programs using coding languages. Even as you perform basic actions like typing the letter A into your document or making a videogame character jump across a fiery pit, your laptop or phone is making that happen by computing those actions in code.

Just as Swedish, Arabic, and English are all complex languages, coding languages can be fairly complex. In fact, for a program to run, it may have to be coded in several languages, one on top of another.

Unlike Swedish, Arabic, and English speakers, though, computers and coding languages are always literal. Computers rely on basic, clear commands that allow them to slowly build up to complex operations. There's no room for judgment or discernment on the part of the computer, and very little room for error on the part of the coder.

Once you master a few of the most useful coding languages, you can perform almost miraculous operations on a computer. Just log into Facebook (or any other popular social media platform, for that matter) and take a look at everything on your screen. The pictures have been coded into the webpage or app. Your ability to scroll has been coded in. Your ability to click and go to different profiles, write text, send links, and use emojis has all been coded into the page. On top of that, Facebook is running highly advanced AI code in the background to choose the best ads to show you. Then, Facebook displays those ads—which are also written in code.

Everything you do on a computer involves code that has been written by some successful software developer, working at a great company, doing some pretty amazing, mind-blowing stuff.

WHY SO MANY LANGUAGES?

If computers can understand code, why not simply invent one language to cover everything? Unfortunately, that's not how computers work. There are hundreds of coding languages, and each language allows a computer to run certain kinds of programs or perform certain kinds of actions. For instance, your Android device runs on Java, while Apple's iOS "speaks" Swift. AI is run on Python and R, while most websites and apps run through C#.

On some level, it would be nice if computers could just "speak" English. That way we wouldn't have to learn coding languages. And the truth is, code does involve a lot of English. C#, for instance, uses "if, then" statements. However, computers can't "speak" English in the way we do. They simply can't process the complex ideas within even fairly simple sentences. Computers require clear commands without ambiguity, and our human languages just aren't good at forming such statements.

Don't worry right now about the differences between those languages. At the moment, all you need to know is that each one is distinct and has its own uses and limitations.

MANY KINDS OF SOFTWARE DEVELOPMENT

Throughout this book, you'll see me using the term "coder" to refer to myself, my Coder Foundry graduates, and others in this field, but "coder" isn't a technical term. It's popular slang. Technically, anyone in this field is a software developer. That's the professional title for someone who writes code either for themselves or for others. Software developers, however, are part of a very broad industry. That job title covers web developers, mobile developers, desktop developers, game developers, AI developers, and more.

Each one of these positions has different responsibilities and specialized skills. For instance, a web developer writes code for web applications and websites that run through a browser. They are responsible for creating Gmail and Facebook as those sites appear in Google Chrome, Firefox, Safari, or Internet Explorer. A mobile developer, on the other hand, writes the code for the Gmail and Facebook apps on your phone or tablet. A desktop developer creates programs like Microsoft Word that run on your computer but not through a browser. Meanwhile, game developers write the code for all the games you love on every platform.

AI is different. In this field, software developers write code that allows computers to review massive amounts of data and "learn" from it. The computer then interprets this data and provides predictions for future actions. AI is increasingly used in almost every industry—from Amazon to Netflix—in order to predict customer behavior and provide better service.

Software development, then, is a huge umbrella term for work across a vast industry. The work a mobile developer does is quite different from that of a web developer, and they are both a million miles from AI. However, all of these careers stem from the same place: coding. So, they are all coders, just as they are all software developers.

Every kind of software developer is needed across the economy, but web development is, by far, the most in-demand position. So while you can become any kind of software developer, in this book, we'll concentrate on making you a web developer, since that's where most of the jobs are. As you'll see, it's also the best place to start your coding career.

THE HISTORY OF CODE

How did we get here? How did the ability to code become the launching pad for success in the twenty-first-century economy? Many of you will remember a time when being on the computer was considered a waste of time. Success, you were told, required learning a more traditionally profitable trade, from factory work to medicine. When did the computer go from a fun device to the key to your prosperity?

Computers and computer coding languages have been around for quite a while. The first coding languages were created all the way back in the 1940s. However, in the decades that followed, these languages remained very specialized. For a long time, computers were massive, warehouse-sized objects with—by today's standards—very limited capabilities. Essentially, they were enormous semiconductor calculators.

This is what a computer used to look like.

Most of these beasts ran on a language called COBOL.
Coders would input the calculations in COBOL, and the
computers would print out the results on green paper.
For large institutions like the military and banks, these
advanced calculations were incredibly useful, but there
was no market for a house-sized advanced calculator for
the small business or individual.

Two things changed computers and the economy forever.
First, the computer shrank in size. From filling multiple
rooms, the computer became a device that could fit on
a desk, and then, in our pocket. Second, two of the most
famous men in recent world history came on the scene in
the 1980s: Bill Gates and Steve Jobs.

In the mid-'80s, Bill Gates made it his mission to put a

computer on everyone's desk. To achieve that goal, he developed Windows, an operating system designed to make computing easy for the average user. Once a computer was small enough and cheap enough for families and small businesses to buy, Windows made it possible for them to use their new devices without a computer science degree.

Once every business had a computer, IT professionals around the world began testing the capabilities of these newly ubiquitous devices. They started creating programs that could make once time-consuming tasks increasingly more efficient. Suddenly, you could print out all of last year's P&L statements in a moment or check month-to-month payroll expenses with the click of a button. By 1991, I was writing code for automobile companies, creating a program that could measure the RPMs of diesel truck engines.

As new versions of Windows continued to make computer use more accessible, the desktop computer became ever more central to our lives. When Microsoft debuted Office—introducing the world to Word, Excel, and PowerPoint—America threw out its last typewriters overnight.

The next major event in the rise of coding occurred in 1997. That was the year Steve Jobs returned to Apple, having been fired from his own company in 1985. He immediately demanded the company "think different," with the goal of making Apple the most relevant company in the world. This new mission led to a string of successes, and in 2007, Apple actually changed history. That's the year Steve Jobs stood on stage and declared everyone in the world was going to want what he had in his hand: the first iPhone.

Just as, twenty years before, every company rushed to create programs for desktop computers, so now every company needed apps for the smartphones in their pockets. Since these phones were not only smaller and more mobile but also more powerful than computers of the early '90s, they could do amazing, unbelievable tasks, so long as someone was able to write the code for it.

We are still discovering the potential within the smart-phone, and businesses continue to race to find the next tech solution to their old problems. At Core Techs, for instance, we've recently written a program that allows workers to "geocode" where cables have been buried underneath the ground. The app can then relocate that cable at any time. Instead of having to do painstaking searches to make sure no one accidentally cuts an underground power cable, workers can open our app and see immediately where every cable is buried in a specific area.

That may seem like a small improvement to you, but it saves power, gas, telephone, internet, and construction companies immense amounts of time and money.

CODE IS EVERYWHERE

At this point, technology has become so pervasive in our lives that coding can be found in almost every device you encounter during the day. Your DirecTV or Roku sitting atop your TV runs on code. So does your DVD and Blu-ray player.

If you have a smart refrigerator, the "smart" side of it is programmed using code. So is any other smart technology, for that matter. Robot vacuum cleaners like the Roomba run the same coding languages as your computer. The key to Peloton's success is the code in the Android app they designed to connect you to all those classes and personalized workouts.

Code is even running parts of your car. This isn't just for those new advanced models that help you brake or stay in your lane. It isn't just Teslas. All modern cars have smart technology that has been coded into the vehicle's computer. Any time your car detects low tire pressure or engine trouble, it's using code. Any time you sync your devices, use GPS, or speak commands, your car understands you thanks to code.

CODING IS CREATIVE

Computer coding is everywhere, and therefore, so are software development jobs. Because every industry runs on code, coders are paid well for their services. However, despite the economic mobilizers, is this really a field you want to join?

There's a stereotype that coders are boring and unsociable individuals who do boring and unsociable work. In almost every television sitcom since the dawn of the desktop computer, the IT person has been looked down upon—a target for jokes instead of a real character.

Believing that stereotype can leave a lasting effect on someone's career path. An ideal candidate for software development may hesitate to learn the necessary skills simply because they've absorbed the idea that the job is undesirable. They wrongly assume the work is tedious and unrewarding—a job for those incapable of developing the social skills to do something more interesting.

That stereotype isn't true at all. While some software developers are introverted, they are by no means unsociable or boring. In fact, when I look at the coders I know, I see some of the most creative people I've ever met. Musicians, painters, writers, filmmakers: thousands of artists make their living as software developers and dedicate their free time to their passion. They make that choice not just because coding pays well, but because the job itself is creatively fulfilling.

Coding is a right-brained activity. That's the creative side of your brain. In fact, the exact same part of your brain that's responsible for painting a masterpiece or writing a song takes control when you're developing a complex piece of code.

This is why musicians often thrive in the world of software development. Developing the code to solve an abstract problem is remarkably similar to transcribing a new musical composition on paper. The same translation from the conceptual to the concrete is required; the same problem-solving instincts are needed.

To those who don't understand the work, it can look like all a coder does is type out formula after formula in weird-looking combinations to get a website or an app to fulfill some basic function. In reality, coding is an art, just like any other. In the world of music, you will find world-class guitar players who make the instrument sound in a way it never has before.

The same creative potential exists in coding. Of course, some basic coding work does involve inputting text that looks like formulas, just like the basics of playing guitar, but when you start trying to make a website or app do something it has never done before, there's a lot of creative exploration that is just as exciting as a new guitar solo.

Think about the vast variety in the look and behavior of the websites you visit every day. Every aspect of those websites comes down to the individual creative choices of a coder. Sure, they are all using the same coding languages, but they are using them in their own unique ways.

In this way, we might compare coding not just to music but to writing a book. If you gave three writers the same outline for a book—in the same language, with the same characters, same plot points, and same conclusion—would you get the same book three times over?

Of course not. Those three writers would make innumerable individual choices, take individuals risks, and explore individual possibilities. In the end, the books may all still be faithful to the original idea, but they'd be completely different stories.

I see this all the time at Coder Foundry. Each student is required to design a bug tracker (a program that tracks coding errors). They all use the same programming languages and have the same goal in mind. Yet, in the five years I've taught this boot camp, I've never seen two students develop the same bug tracker. And I never expect to.

Each bug tracker accomplishes the same goals, but each one looks and behaves differently. Each one was the creative effort of an individual coder, and the individual creative effort always shows through.

ARE YOU A CODER?

Software development is creative, high-paying, and in-demand across the country. It's a highly desirable job. The only question left, then, is whether this is the job for you.

The best way to find out is to take our free Software Developer Career Quiz at www.coderfoundry.com/quiz. This sixty-minute quiz tests if you think in the way that coders have to think: logically, creatively, and pattern-based. If you score above eighty on the quiz, you definitely have what it takes to become a software developer. You can't start learning to code fast enough.

If you score fifty or below, there's nothing wrong with you. You just think in a different way than most coders. That's okay, but it means that coding may not come to you naturally. That doesn't mean you can't learn it, but you may have to work harder. If you choose to pursue a software development career, you should keep in mind that you may often have to put more energy into learning and creating code. This can still be a great opportunity for you—the money and the potential will be the same—but it will require more time and commitment.

Whatever your score, if you want the life that coding can provide, it's there for the taking. All you have to do is learn how to reach that promise. As we're about to find out, the process can take far less time than you might assume. In fact, you can get from today to your first software development job in just five steps.

PART II

THE FIVE STEPS TO A CODING CAREER

—

{ CHAPTER 3 }

STEP 1: LEARN THE RIGHT STACK

I'm very active on social media. It's how I engage with seasoned and novice developers around the world. Amongst the thousands of posts I see every week, there are always a handful from people complaining that no one is hiring software developers anymore.

Here is a perfect example I found on LinkedIn from a man named Craig:

"Do people even hire software developers anymore? Every time I turn around, it looks like employers only want C# or .NET developers. I don't know if I can get a job with the languages I know. I should have learned something else."

Craig made a classic mistake when learning to code: he

didn't learn the coding languages that lead to jobs. Unfortunately, many new coders make that mistake. There are hundreds of coding languages, thousands of schools that teach how to code, and tens of thousands of software development jobs. With all that variety, it's very easy to approach coding with all the right intentions and plenty of motivation and still fail to get into the industry.

There is a clear pathway that leads from technology neophyte to a six-figures-a-year, expert software developer. That path involves your choice of educational institution, your portfolio of programs, your interview technique, and your work with a recruiter. But it starts with learning the right coding language stack.

WHAT IS A STACK?

Since every computer language has fairly specific uses and limitations, you can't really learn just one. Instead, you have to learn a stack, which is a collection of related coding languages and tools. A language stack will cover all the coding requirements to create certain apps and programs. You can think of a stack like a toolbox. One language might be a hammer. That's very useful in some situations, but you also need screwdrivers, tape measures, pliers, and saws to do a serious construction job.

There are many different stacks out there, but by far, the

most useful, most popular, and most employable stack is one that Craig mentioned in his LinkedIn post. The .NET stack allows you to design web applications, which, as mentioned in chapter 2, is the most in-demand position in America. While there are plenty of interesting stacks available to coders, .NET is the first one you should learn if you want a career in this industry.

ASK A BUSINESS, NOT A CODER

Many people, when they decide to learn to code, ask someone they know in the business for advice on what language to learn first. It's a reasonable strategy. If I wanted to know the best way to start studying medicine, I'd probably ask my doctor. If I wanted to learn how to teach, I'd ask a local teacher.

Software development is a different kind of industry. Asking a software developer what coding languages to learn is closer to asking an artist what techniques to learn to become a painter. Every artist is going to have a different answer.

At the same time, coders are extremely fickle. If people married and divorced as often as software developers changed their coding language preferences, they'd be in for a very turbulent and unhappy life. Coders can fall in and out of love with languages on almost a weekly basis.

Not so long ago, one of my developers, who I'll call Jeremy here, insisted on coding a project with a framework called Knockout. He told me that it was the greatest framework out there. He was so enthusiastic, I gave him the go-ahead. A few months later, Jeremy got his opportunity when we started a new project that required a new framework.

"Didn't you say Knockout was the perfect framework for our projects?" I asked him.

"Knockout sucks," he told me. "We need to use Angular now."

In a matter of months, Knockout went from the greatest coding framework ever developed to a framework he couldn't even stomach for one project.

Software developers never stop learning new languages and frameworks, and frameworks and languages are constantly going in and out of popularity. Coders always have a new language that excites their curiosity. That's a great instinct to develop once you have a few years' experience in the industry and have the room to specialize, but it doesn't make for very good advice for beginners.

So, when you are starting out and want to know which stack to learn, don't ask a coder what languages they think are cool; look at what languages businesses need you to learn. Across the board, the answer will be the same: .NET.

THE .NET STACK COMPONENTS

Now, don't get me wrong. .NET is cool, and you can do a lot of cool things with it. It has just been around for a long time, so many coders gravitate toward what is new instead of what is best.

The .NET stack is the best place to start because it is the most widely used stack across the whole economy, and that makes it the one that will get you a job as soon as you learn it. But what is it?

The .NET stack for web development includes several languages (C#, HTML, CSS, JavaScript, and SQL), a design pattern (MVC), and a program (Visual Studio). The .NET stack is an incredibly flexible, useful set of tools that go far beyond just web development, but for the purposes of getting a job, we are going to focus on its web development capabilities here.

The whole of .NET revolves around the first of those languages, C#. C# runs what is commonly called server-side code. This is the language that allows your computer to connect through your browser to a server—which holds all the information about a website—and load that website on your device. Once the server and the computer "speak" in C# together, the website's information is displayed using other languages, such as HTML, CSS, and JavaScript. These languages allow the website to show text, images, and video. They also allow you to click links, search, and scroll the page.

Finally, if the server needs to connect to a database for certain information about the user—such as account details—it will use SQL.

With this combination of coding languages, a software developer can create any of the useful and dynamic websites you frequent every day.

We can see this by considering a single site. Let's say I want to look at my account balance on Duke Energy's webpage. I open my browser—Chrome, Safari, Firefox, etc.—and type in "duke-energy.com." That address is a request from my computer to the Duke Energy server, and that request is made in C#. Once the request has gone through, the Duke Energy page opens using HTML, CSS, and JavaScript for all the various page elements.

To get any further than the Duke Energy's homepage, I have to log in. So, I put in my information, and the Duke Energy server checks my name and password using C# again. Once it confirms my identity, a new page opens with new HTML, CSS, and JavaScript elements.

At this point, I want to see my balance, but that balance isn't on the server, it's in a database written in SQL. Using

SQL, Duke Energy's server communicates with the database and then communicates the information back to my computer using C#.

This whole interaction takes a few seconds and requires almost zero effort on my part, but every step of the process has been thoroughly coded by Duke Energy software developers using the .NET stack of languages.

To write in these languages, you need a design pattern to make them easy to read. A coding design pattern is like the old essay pattern we all learned in school: introduction, body, conclusion. Each idea gets a paragraph with an introductory sentence and a concluding sentence. In the same way, a coding design pattern breaks up elements in the code so they are easy to read, and the logic is easy to follow.

This pattern is called Model-View-Controller, or MVC. It exists solely to help coders create multiple projects using the same pattern or approach. A computer will understand code written in many different patterns, but we want to make sure our fellow software developers can read our code, understand it, and make changes.

You can think about MVC like architectural blueprints. You could design a great house using your own system of measurements and sketches, but you couldn't expect the construction team building the house to implement your plans without great effort. To avoid unnecessary complication and potential mistakes, architects and construction workers have an agreed-upon system that they all understand. The same is true of coders. MVC is universally accepted by the whole community.

The last piece of .NET is the Visual Studio program. This program is called an IDE, which stands for "integrated development environment." An IDE acts as your code editor, debugger, and sandbox, where you can play around with potential coding ideas. It is like an artist's practice canvas. You can experiment here, make mistakes, see what works, and fix what doesn't. There are plenty of IDE programs like Visual Studio on the market, but Microsoft designed this program specifically for .NET.

WHY THE .NET STACK IS KING

.NET is the number one choice for almost every business with a web development shop because it's backed up by one of the kings of the whole tech industry: Microsoft. Over the past twenty years, Microsoft has invested $2 billion in marketing .NET to developers. At this point, corporations have already spent the first two decades of the twenty-first century adopting .NET and integrating that infrastructure into their systems.

Because it is so entrenched in businesses across the

planet, .NET is, essentially, the lingua franca of the web development world. Almost every coder can write in these languages and almost every company uses these languages, even if they also work with other stacks.

In the world of web development, .NET is as universal as the English language in the world of business. Multinational businesses around the world work and advertise in English. Most of the world's biggest markets speak English as their first or second language. At the same time, to negotiate with other international businesses, it's just easier to speak in English, since the employees of most companies will almost certainly be fluent in the language. Since the international business world already uses English, new corporations introduce the language in their offices. Again, it's just easier to work with the system that's already in place and already successful.

Just like English in the office, .NET is the stack of choice worldwide. It has proven to be extremely successful, and very few businesses will choose to code their websites and web apps with any stack other than .NET.

Even if a business wanted to introduce a new stack into their web development shop, the price of doing so would be prohibitive. Businesses spend millions of dollars over several years to work with the .NET stack because of its primacy in software development. They aren't going to turn around and spend millions more just because .NET seems less fashionable.

In fact, many banks still use COBOL—a language developed in the '50s and '60s—as the base language of their programs. They simply have software developers write .NET code on top of their COBOL applications because they can't justify the cost involved in fully updating the system.

LEARN THE COOL STACKS LATER

Just because you're learning the .NET stack first doesn't mean you are tied to those languages and web development forever. .NET is just the easiest way to break into the industry. Once you're in, you can learn any language you want and pursue any software development career.

But you have to start with .NET. Once you know .NET, you immediately make yourself employable. Within days of finishing a .NET course, you can start earning a great income. Then, you pick up the rest. If you learn Rust first, you might not be able to find an entry-level software development job. Then, you have to put more time and money into learning .NET on top of the time you spent learning Rust. So, just learn .NET first, and you can get paid to learn Rust.

Once you've proven you can work in the coding world with

.NET, businesses will trust you can learn whatever else is required later.

In the '90s, one of the more popular languages around was PowerBuilder. So, when I sat down for one of my interviews, the development manager naturally asked me if I knew it.

"No," I said.

"Can you learn it?"

"Sure."

And that was it. I got the job. Because I had already proven that I could code, the new language wasn't a barrier. That company was willing to pay me while I learned the coding language they really needed from me.

That's the difference between successful coders like Jeremy and unsuccessful coders like Craig. Jeremy learned the basics, so he can learn the newest, coolest languages now while he makes a great income. Craig learned the cool stuff first, and he's still looking for work.

So, you have to learn .NET. Now the question becomes where to learn it. Many educational institutions promise to teach you how to code, but unfortunately, most of them fail to live up to that promise. Luckily, there is one option out there with a history of successfully placing new coders in lucrative positions: the in-person boot camp.

{ CHAPTER 4 }

STEP 2: CHOOSE YOUR SCHOOL

Patrick Richardson knew from an early age that he wanted to work in software development, so he pursued his chosen career in the traditional way. He went to UNC Charlotte and earned a four-year computer science (CS) degree. Like most college students these days, he spent a lot of money, took on a lot of debt, and expected that investment to pay off.

It didn't. Despite his hard work and good grades, he found himself in the same position Craig did in the last chapter: unemployed.

After six months of looking, Richardson came to Coder Foundry. I took one look at his degree and asked the obvious question, "Why are you here?"

His answer was shocking.

"I didn't really learn to code in college."

Despite four expensive years of higher education supposedly concentrated on computer science, Richardson's college hadn't actually taught him the practical skills he needed to enter the world of software development. Instead, he needed to enroll in our twelve-week boot camp to develop his coding knowledge enough to get his first job.

Richardson did everything right, yet he had to pay twice to learn how to code once. This is the sad reality behind software development education. While many careers require a four-year degree from a respectable university, as Richardson found out, software development is different. In this sphere, there are clearly better, cheaper, and faster ways to become a coder than the local university.

THE LIMITATIONS OF TRADITIONAL EDUCATION

Broadly speaking, there are four ways you can learn how to code:

→ A four-year university degree
→ A two-year technical/community college degree
→ Indefinite DIY online boot camps and books
→ A twelve-week intensive in-person boot camp

Most of us are told from an early age that the first option—a four-year bachelor's degree from a university—should be the best. Throughout our middle school and high school academic careers, we're encouraged over and over again to apply for and complete a four-year degree at a reputable school.

Under most circumstances, I have nothing against the university system in America. For many careers, that is the right choice. I certainly wouldn't hire a lawyer who pieced together their education on their own or go to a doctor who didn't complete all those years of postsecondary medical education.

However, the university system in America has yet to develop a system to effectively teach coding. There are three major reasons for this.

The first is a lack of focus on coding languages and skills. To be fair, Richardson's education—and most CS degrees—include some coding classes, but the skills acquired in those classes are never brought together and connected to full-stack coding projects. So, while a student like Richardson may learn C# in one class and HTML in another, they are never taught to connect those languages and use them to build a website or an app.

Instead, universities try to teach a little of everything over the course of a CS degree. There may be a class on C#, but there will also be a class on databases, one on servers, and another on statistics. All of these classes are interesting and informative, but they don't teach many skills that can be applied to a job later. By the time a student graduates, their C# skills learned sophomore year are extremely rusty, and they have no idea how to use those skills in a real job developing real projects for their employer.

That lack of interest in real-world application is the second problem. Professors are extremely intelligent and knowledgeable about computer systems, software, and coding languages, but they become professors because they love knowledge for its own sake. And that is often how they teach. They aren't interested in whether their classes teach applicable skills. They're interested in the theory behind those skills.

A lot of times, university instructors have never coded for a living. So, when they teach a class on HTML, they aren't going to teach you how to use HTML to build a fully functional web application for your future employer. They're going to teach you the theory behind HTML and how professors are using the language to advance academic study. In other words, a CS education at a four-year university focuses on concepts and theories, not practical skills useful to a career in software development.

Even if the instructors at your university were focused on

practical skills, there would still be a third, significant problem with the teaching process. Software development is a constantly evolving industry. Just think of the apps you were using five years ago and how they differ from the apps you use today. In order to keep up with the changes in the design and function of websites and apps, software developers need to know the latest languages and techniques.

That need conflicts with the process behind university course certification. Universities spend years agreeing to their course curriculum. By the time they have created their most up-to-date class on coding, they are years behind the marketplace.

The combination of outdated materials and out-of-touch instructors means universities often produce graduates like Richardson who don't know how to code. And perhaps worst of all, the universities themselves aren't even aware they're doing anything wrong.

Recently, I was talking to a friend of mine who is a professor at a prestigious four-year university. We were having lunch together, and our conversation turned to the differences between how his university and my boot camp taught students. He started making fun of Coder Foundry, saying we never did anything interesting or fun in our classes.

"So, what's the last project you guys completed?" I asked.

"Well," he told me, "we made an iPad into a web server, and now we're hosting websites on an iPad."

"But that's useless," I said. "What's the point of doing that? That will never work in production. No business can use that. How will that help your students get a job?"

"It doesn't matter. It's cool," he objected.

"That's the real difference between us," I said. "At Coder Foundry, we teach students to build projects that real companies need. You build projects because they're cool."

To be fair to him, I'm sure his students had a lot of fun on that project, but he still did them a disservice by spending valuable teaching time creating something that gets those students no closer to a future career in the field.

Despite my arguments, though, he couldn't understand my perspective. To him, coding is purely conceptual. It's about the ideas behind the computer languages.

I'm sure academics like my friend are coming up with amazing new ideas every day. However, unless you intend to be a CS professor yourself, that four-year CS degree is not going to be particularly useful. And it comes at a high price in terms of time and money.

COMMUNITY COLLEGE HAS UNIVERSITY PROBLEMS

You can see how ineffective four-year CS degrees are for coders by their low completion rate. At the moment, only 33 percent of students complete their four-year CS degree within the expected timeframe. Two out of three CS students either give up, dedicate even more years (and money) to the program, or find a better way to learn.[5]

The reasons for this are obvious. Beyond the systemic limitations of CS degree programs, universities also require four years of your time and tens of thousands of dollars. Over that time, life often gets in the way. People run out of money or lose motivation. Even when they make it to the end of the program, they find they aren't prepared for the software development job market.

I'd love to tell you that community college is a better option, but unfortunately, they also only graduate 36 percent of their students into careers.

Despite the same graduate rate, though, there are some advantages to community college. Compared to a four-year university, a two-year college degree is obviously cheaper and requires half the time commitment. Two-year colleges also focus more on practical skills. They are more

---

5   Lynn O'Shaughnessy, "Federal Government Publishes More Complete Graduation Rate Data," *College Insider with Lynn O'Shaughnessy, Cappex,* https://www.cappex.com/articles/blog/government-publishes-graduation-rate-data.

concerned with teaching actual coding languages instead of the conceptual framework behind how computers work.

That all sounds pretty good, but community colleges still suffer from many of the same flaws as four-year universities. Like universities, the community college certification system causes courses to lag behind the industry. Going to a two-year college, you'll learn about older languages that are no longer entirely applicable to today's businesses. Also like universities, most instructors lack real-world coding experience. So, they are ill-equipped to help you prepare for the complicated software development interview process.

In this sense, the community college is still more academic than career-oriented. The end goal is still the education. At a boot camp, on the other hand, the end goal is a job. That remains the major difference.

THE DO-IT-YOURSELF ONLINE BOOT CAMP

Unfortunately, because American educational facilities have failed to adequately teach coding skills, the vast majority of coders—69 percent—are at least partially self-taught.[6]

In the old days, before we had so many online resources,

6   Michael J. Coren, "Two out of Three Developers Are Self-Taught, and Other Trends from a Survey of 56,033 Coders," *Ask a Developer, Quartz*, March 30, 2016, https://qz.com/649409/two-out-of-three-developers-are-self-taught-and-other-trends-from-a-survey-of-56033-developers/.

most of us learned through books, friends, and messing around in Visual Studio after school. Today, there's a whole industry designed to help potential software developers learn their trade on their own: the online boot camp.

These boot camps provide online courses that are meant to guide "students" (I use the term very loosely here) to coding excellence. That sounds great—in theory.

The truth is, these programs are not cheap—they often cost as much as in-person boot camps—and they have serious deficiencies.

Those deficiencies begin with the lack of an actual person in charge of your studies. For all their failings, universities and community colleges always provide you with instructors who guide your work and review your progress. With online boot camps, all you have are prerecorded videos.

Some of these programs try to compensate for this issue by providing online teaching assistants. However, these teaching assistants are often not experts; they're simply ahead of you in the program. They may not have even finished the same course you're taking.

That's only the beginning of the trouble. Because these programs are one-size-fits-all and lack instructors, the content often makes assumptions about the knowledge a student brings to the course. Far too often, online boot camp classes will assume the viewer knows certain background information about coding. In a normal class, a student can ask their instructor to clarify a point. With online boot camps, they won't even know that they're missing crucial techniques until they try to replicate the instructions, perhaps months later.

And then there's the time requirement. Many of these programs require at least nine months. That's less time than a community college or university degree, but as we will see shortly, it is far longer than in-person boot camps. Other programs have no time constraints at all. They are learn-at-your-own-pace programs. That might sound appealing, but it increases the chance a student will lose motivation and slowly drift away from study.

All of these factors contribute to the extraordinarily low completion rate for online boot camps. Only 6 percent of students finish these programs, and honestly, most of those students would have been better served saving their money and studying coding books.

About 10 percent of all software developers are gifted and motivated enough in this area that they could teach themselves to code and get a job without any assistance. For those people, online boot camps are a waste of money. For the rest of us, these programs have discouragingly poor results.

If you think you are dedicated and talented enough to complete an online boot camp, put this book down, go buy a set of coding books and Visual Studio, and teach yourself the way we learned in the old days. Otherwise, take your online boot camp money and put into a program with far better results.

IN-PERSON BOOT CAMPS ARE BETTER

The four-year university, two-year community college, and online boot camp all have glaring limitations. They cost a lot of money, take a lot of time, and have discouragingly poor records placing graduates into the workforce.

But don't feel discouraged. If those three education options don't appeal to you, there is a final, far superior way to learn to code and find success in software development: the in-person, immersive boot camp.

In-person boot camps are almost always the cheapest, fastest, and most effective way to learn to code. Coder Foundry, for instance, follows the industry standard and provides a twelve-week intensive course. Our students pay about the same amount they would to "attend" an online boot camp, but they complete the course in a third of the time.

The in-person boot camp is designed to address the objections I've raised about software development education. To begin with, these boot camps are firmly focused on practical skills. At Coder Foundry, we reverse engineered our whole course to ensure job placement. We started with the ultimate goal of landing that first, high-paying web development job, and then we worked backward, making sure we taught everything a student needed to know to go from complete technological ignorance to coding success story.

Almost as important as the bread-and-butter issues of time, money, and practical use, the in-person boot camp offers something no other educational experience can: a real mentor experienced in the industry. Most reputable, in-person boot camps only hire instructors who have extensive experience working as software developers in the real world. These instructors know what it takes to get into the industry and how to advance because they have done it themselves. They know the practical pitfalls that often arise when learning to code, and they are always available over that three-month period to address questions or weak points in their students' knowledge.

These instructors can also provide the kind of motivation you need to take in so much knowledge over a short period of time. Coding isn't hard; the complexity of the work is part of the attraction. However, because software development can be such a difficult skill to master, you need an instructor who is more than a source of knowledge. In reality, you need a coach.

All successful sports teams need a coach who knows not just the sport but how to inspire the players when the going gets tough. When the team is down and time is running out, the coach has to be able to motivate every player to dig in, find that last bit of energy, and commit it to the team's success. That's precisely what a boot camp instructor can provide you.

At Coder Foundry, we talk to every one of our students at least once a week, and we work through how they're doing, how they're feeling, and how they can push through the struggle to reach success. Nothing inspires a student feeling down about themselves like a hand on their shoulder and an experienced voice telling them, "You can do this. Don't quit now."

That's why so few people quit in-person boot camps. That's why we have an 85 percent success rate compared to online boot camp students who succeed only 6 percent of the time.

No other software development educational program can provide that kind of mentoring experience. Online boot camps provide no mentor at all. All the motivation has to come from you. Even at the university or college level, the mentor system isn't as effective. University or community college professors can certainly be mentors, but they lack the real-world experience to properly motivate and guide their students toward success.

The only program where you will receive inspirational, personal, and experienced instruction is the in-person boot camp.

INSPIRED BY REAL BOOT CAMPS

There's a reason we call our course a "boot camp." The course is designed to mimic the successful military system.

When a person joins the military, he is sent to boot camp where a drill sergeant is assigned to train him for combat fitness. The drill sergeant is an experienced soldier who knows everything about the battlefield. Their purpose is to give the recruit instruction, guidance, and motivation, so he, too, can be successful in a combat zone.

The entire system is designed to ensure each soldier receives all the information and feels all the motivation they need to protect themselves and their fellow soldiers. With lives on the line, the system has to be as close to perfect as possible.

Consider the SEAL boot camp. Navy SEALs are among the most elite military men and women in the world, so their training has to be the best. In SEAL boot camp, the recruit has to ring a bell to demonstrate to everyone he feels he can't complete the training. The SEALs provide that social pressure because they want to use the strong bonds the recruit has forged with his comrades in arms to push himself a little harder and go a little longer. They don't want the recruit to quit. They want to give him the strong incentives he needs to succeed.

The boot camp environment inspires people to accomplish amazing things. They run that extra mile and complete that ninety-eighth push-up. They reach heights beyond their wildest dreams.

At Coder Foundry and other in-person coding boot camps, we aren't going to make you run anywhere or do any push-ups, but we will push you to be all you can be as a software developer.

LEARNING A NEW LANGUAGE

On many levels, the process of learning a coding language is very similar to learning a foreign language. Intensive, immersive training works best in both cases.

Let's imagine you've decided to learn Spanish, and you have four different educational options. Option one is to enroll in a university and take four years of Spanish classes that focus on grammar and the history of the language more than speaking it to native speakers. Option two involves a two-year course with more speaking classes but essentially the same system.

If you've ever taken academic language classes, you know how limited the educational benefits are for either of those options. Some people do gain fluency by the time they earn their degrees, but most students end up with a limited ability to speak the language.

So, that brings us to option three: online language classes. With this system, you can watch YouTube videos and take online Spanish language quizzes as often as you want for as long as you want.

At this point, how likely do you think it is that option three will actually transform you into a Spanish speaker? Sure, a few people do learn the language that way, but they are exceptions, not the rule.

Instead, to truly learn Spanish, most people will need some version of option four: immersive study abroad.

If I dropped you off in the middle of Mexico with a teacher who speaks the language, you'd be fluent in Spanish in three months.

Of course, you'd need some translation help along the way from your teacher, but the immersive environment would do a lot of the work for you. You wouldn't be able to order lunch or dinner without speaking Spanish. So, you'd have no choice but to become more comfortable with the language. The motivation would be in the environment, and the constant practice would improve your skills quickly.

An immersive language learning experience is always better than a standard academic or online classroom experience. You're surrounded by people who know the language and who are learning the language. Your days are dedicated to mastery of that language. It's the same whether you're learning Spanish in Mexico or C# at Coder Foundry.

THE GOAL: GET A JOB, NOT AN EDUCATION

At Coder Foundry, we don't consider ourselves to be in the education business. Instead, we're in the business of changing lives, and nothing changes a person's life like getting a well-paying job that puts them on track for a financially, professionally, and creatively successful career.

Patrick Richardson told me after he finished the Coder Foundry course that he learned more in twelve weeks than he had learned in four years at a university. He's now three years into a successful career, just like 85 percent of his fellow Coder Foundry graduates. The majority of our current students already have jobs lined up with salaries above $50,000 a year. They can start the day after they graduate.

Part of that success comes from the boot camp environment, but part of it is about something special in our approach to education. We don't just teach you how to code. We teach you how to get a job. That education involves the necessary coding skills, but it also involves teaching you how to navigate the interview process, which is more complicated than in many other fields.

This part of your education involves working with recruiters and fine-tuning your interview technique, but it starts with building a portfolio that can earn you instant credibility with all your potential future employers.

CHAPTER 5

STEP 3: BUILD A PORTFOLIO

The HBO sitcom *Silicon Valley* is about a group of bumbling software developers trying (and often failing) to make it big in the tech industry. One episode involved the absurd premise of an app that could identify what kind of food someone was eating. One of the coders, Jian-Yang, created just such an app. The only hitch: the app could only say if the food was a hot dog or "not hot dog."

It's a funny scene in a very funny show, but it's not good advice for new software developers creating a portfolio. A coding portfolio is a collection of the apps and websites you have created that demonstrate your skill. Unfortunately, because so few schools and online boot camps teach coders how to build a portfolio, many novice coders don't understand why a hot dog app is inappropriate for an interview.

One such coder, Sarah, actually built that exact app and made it the centerpiece of her portfolio. Like Craig in chapter 3, she posted online, asking others to look at her portfolio because she was mystified why she couldn't find a job after finishing her boot camp.

I looked through the rest of her portfolio, and I had to break the bad news to her: she was never going to get a job in this industry with the projects in her portfolio. She'd have to start from scratch.

The joke in that *Silicon Valley* episode is that an "Is It a Hot Dog?" app would be completely useless. It's a funny gag, but in real life, a development manager isn't laughing when you present that app to them. You can try to spin that project as a sign that you can do anything they want, but a development manager has a specific, real need at their organization, and they aren't interested in amusing hypotheticals. When they look at a portfolio, they want to see projects built to solve their real business problems.

KNOWING HOW TO CODE ISN'T ENOUGH

The one thing Sarah got right in her interview preparation was the need for a portfolio. Without a portfolio, you have no proof that you are as good at coding as you claim to be, which immediately puts you at a disadvantage. The interviewer has to resort to asking you "code trivia questions" to see if you can do the job.

These random, complex questions test how much you know about coding languages and techniques. Your interviewer will expect textbook answers to their code trivia questions. If you forget anything, you'll miss out on the job.

With a portfolio, on the other hand, you can show your interviewer just what you're capable of. The interview will be about the programs you have developed, not random trivia.

A portfolio, then, is an incredible advantage when breaking into this industry, and yet, most new coders don't have one when starting out.

That fact should encourage you. It means you can distinguish yourself from the competition. When you walk into the development manager's office with a portfolio, you communicate that you're motivated, capable, and already experienced in creating a professional final product.

The interviewer won't have to quiz you on code trivia to find out if you can solve their problems. You'll have already proved it.

The fact is, though, that all those advantages in the interview disappear if your portfolio is stuffed with useless, silly programs.

A GOOD PORTFOLIO HAS THE RIGHT KIND OF PROJECTS

Sarah's "Is It a Hot Dog?" app was a particularly bad choice for a portfolio, but I've seen plenty of other poor selections. Many new portfolios will include fun but pointless projects like tic-tac-toe and weather apps. Unfortunately, even some in-person boot camps will encourage giving these creations pride of place in a portfolio.

On a very basic business level, these projects are as useless to a company as turning an iPad into a server, like my friend the professor did. Some people assume these projects speak well of them. They show a sense of humor and creativity. But that isn't what most interviewers see. They see a person who wastes time on useless projects; someone who would rather make something fun than something purposeful. They don't see a charming, creative person; they see a questionable investment.

You may be able to convince them otherwise in the interview, but why make the process harder on yourself? If your portfolio already shows you can make high-quality, professional apps, you don't have to prove anything.

I don't mean to disparage fun, creative coding. By all means, make silly apps in your free time. There are jobs out there that will respond positively to your creative spirit. However, those jobs aren't usually available to recent graduates. You have to prove you can do the serious work first before you gain access to the coding playground.

So, prove you can do the serious work by using your portfolio projects to demonstrate the skills most businesses are looking for. When a company hires a software developer, they need that person to create code that accomplishes five things:

→ Addresses a business problem
→ Establishes security measures that ensure only the right people can access and make changes to information
→ Uses a design pattern that every coder can understand
→ Connects app data to a database
→ Improves communication within and outside the app

Every single project in your portfolio should demonstrate your ability to address all five of these criteria. That way, when the development manager reviews your portfolio, every project shows them that you are capable of fulfilling the requirements of the position.

DRAWING SPIDER-MAN FOR MARVEL

You want your projects to highlight the five criteria above, but what should those projects actually look like?

In a word, Spider-Man. That's the advice comics great Todd McFarlane gave to new artists trying to get into the comic book profession.

Many artists send drawings of their original characters to big comic book companies like Marvel or DC in order to show how talented they are. McFarlane says that is the wrong approach. Instead, artists should send Marvel their best picture of Spider-Man, and they should send DC their best picture of Batman. Why? Because the people at Marvel know what a good Spider-Man looks like, and DC knows a good Batman when they see it. They've seen a hundred artists draw those characters, and they know when it's done well.

Can you tell if this is a good Spider-Man or not?

It's the same in software development. You want to create programs that a development manager can immediately recognize as useful to their organization. That way, they can focus on the details of each project that show just how good you are.

CODER FOUNDRY'S THREE PROJECTS

At Coder Foundry, we concentrate on developing the kind

of portfolio projects that impress development managers. In our twelve-week course, every student builds three projects: a blog, a bug tracker, and a budgeting app. Each of these apps will be immediately recognizable to an interviewing development manager. And each app addresses all five criteria laid out above—solving a business problem, establishing security, using a design pattern, connecting to a database, and improving communication.

Obviously, we always use MVC, so the interviewer will recognize the universal design pattern right away. Each project also stores data in a database, demonstrating that capability as well.

However, each project addresses the other three criteria a little differently.

THE BLOG

Our first portfolio project is a blog. This isn't just a writing assignment; our students construct a whole blogging site on which to record their coding progress. They develop the frontend (the part of the site visitors see) and the backend (where all the technical coding work takes place). They also create a security system that allows the blog owner to determine who can author posts and who can make changes to the site.

This blog is useful on many levels. Most importantly, it is

immediately recognizable as a useful project to the development manager. Every business needs a blog these days, so with this project, the interviewer will have an excellent example of how your coding can provide a solution to a real-world business problem. They'll know they can trust you to build an attractive, functional, and secure website. You'll have proven you can create a powerful communication tool for the business and that you can communicate effectively on that platform.

Additionally, if your interviewer glances through the posts, they'll see updates that cover all of the languages and skills you've developed over the course of your boot camp.

In short, this blog is a massive sales tool. With this one project, you can promote yourself as an expert—both in terms of words on screen and in terms of coding skills—and demonstrate how useful you will be for your new employer.

THE BUG TRACKER

A bug tracker has nothing to do with insects flying around the office. Instead, a "bug" in the software development world is a mistake or defect in the code that causes an app to malfunction. If you've spent any time at all on a computer or a smartphone, you've seen bugs, even if you didn't know it at the time. Have you ever clicked a button on a website and it didn't work? That was a bug in the code. Have you ever opened an app only for it to immediately crash? That was another bug.

Unless you catch these mistakes immediately, it is very hard to find a bug in the thousands of lines of code required for an app to run. That's where a bug tracker comes in. A bug tracker organizes a team around correcting mistakes. First, it logs defects reported by users. Then, it allows the project manager to assign a defect to a specific developer. Once the developer says the bug is fixed, a QA (quality assurance) tester checks the work. After they sign off on the corrections, the project manager publishes the new code.

This is obviously a very complex process, so a good bug tracker solves a very serious and significant business problem. But to be a good bug tracker, it has to do more than just organize the code-fixing process; it also has to provide security by restricting what each person can do (so a QA tester can't fix an error and a developer can't publish a change) and create a framework through which each person on the team can communicate and report their progress.

Since bug trackers are so important to software development, any interviewer will know what one looks like. They'll also know if yours works, and if it is as effective in accomplishing our five criteria as the program they currently use. If it happens to be more effective than their current program, you can offer your app as an upgrade that comes with hiring you.

I've seen this upgrade happen in interviews. One of our former Coder Foundry students, Ryan, interviewed for a job at a development shop that didn't currently have a high-quality bug tracker. After the interview, the development manager approached me and whispered conspiratorially, "Can we use this app if we hire Ryan?"

"Of course," I told him.

Needless to say, Ryan got the job, and the company implemented his software on the spot.

THE BUDGETING APP

You've probably seen budgeting apps before. You may even have used one. These apps are an incredibly popular way to keep track of income and expenses for individuals and families. Every Coder Foundry graduate builds just such an app for their portfolio.

At first glance, a budgeting app may seem no more useful to a business than tic-tac-toe, but a development manager can immediately recognize how a budgeting app should work and how the skills required to build such an app could be used in a business context. After all, many businesses need their software developers to create apps that are easy to understand and easy to use for nontechnical colleagues. Having someone in-house who has proven they can develop such apps is incredibly useful.

At the same time, a budgeting app also addresses security concerns by keeping data and communication separate for different households. Those same tools can be used in apps that store data for multiple teams, all with boundaries installed to ensure only the right people have access to the data.

And since our budgeting app allows multiple users to work within a single household, it enables quick and clear communication between all parties.

So, while this app won't be immediately used by your new employer to upgrade their own apps, it still shows all the skills you bring to their organization. That makes it far more useful than tic-tac-toe.

JUDGING BETWEEN PORTFOLIOS

If you want to learn more about our portfolio projects, we've set up a site—FiveStepsToASoftwareJob.com—where we can demo each one and show precisely why they catch the eye of interviewers.

Of course, these aren't the only projects you can create for a great portfolio, but they are representative of the types of projects you should pursue. Any project you put in your portfolio should address all five of our criteria and be immediately recognizable to the development manager interviewing you.

When you bring a portfolio with programs like that to an interview, you immediately place yourself in pole position for the job. Just think about it. Who do you think a development manager is going hire among the following job candidates?

→ You, a Coder Foundry graduate with a practical portfolio
→ Someone with just a resume, but who graduated from a prestigious school
→ Sarah with her "Is It a Hotdog?" app

The answer is obvious. You, the boot camp graduate, are the best choice because your skills aren't academic—they're real, demonstrable, and practical.

MAKE YOUR PROJECTS ATTRACTIVE

Before we move on, I have one final tip to really make your portfolio a winner: make it pretty.

People love attractive products. That's true across the whole spectrum of the human experience. If something looks good on the outside, people will assume what's inside is equally good.

If you put a Mercedes engine into a broken-down Ford, people wouldn't bother to check what's under the hood. But if you put a Ford engine into a Mercedes frame, most people couldn't care less whether the engine matched the outward design.

The same principle holds for software development. You want your portfolio to be functional, but you also want it to look good. Whatever apps you put in there, they should look like Mercedes, not broken-down Fords.

When you're competing for a position against other software developers, the fact that your portfolio looks the best can make all the difference. Make the quality obvious from the first glance. That's what gets you hired.

Essentially, you have to be a great salesperson for your coding skills. Part of that salesmanship is putting the right projects in your portfolio. Part of it is making those projects attractive. And part of it is knowing how to sell yourself and your projects to a development manager once you're finally in the room with them.

Just as that Mercedes sitting in the car lot isn't going to sell itself, so too your portfolio only gets you so far. Once you have the projects, you have to know how to talk about your code.

CHAPTER 6

STEP 4: TALK ABOUT YOUR CODE

My Core Techs partner, Dave DeBald, and I have been working at the highest levels of the software development world for almost thirty years. So, you'd expect us to know how to answer any question that might come up in an interview. And, usually, you'd be right.

A few years ago, though, Dave had a "brain blank" moment. He was interviewing for a consulting gig with a company called Inmar. They handle most of the coupon redemptions for grocery stores across the country. The interview started well, but instead of focusing on Dave's successes, the interviewer started to play the "code trivia game," testing Dave's knowledge of obscure coding lingo. She asked Dave a question, and Dave blanked on the answer.

Now, Dave is, by far, the best coder I've ever worked with. He's one of the smartest people I've ever met. Yet, for some reason, this one code trivia question tripped him up. Despite all of his experience and his otherwise excellent answers, missing that one question cost us the contract. The company hired someone else to do the work, and we moved on to the next client.

However, as is often the case with those "code trivia game" interviewers, they hired poorly. It often turns out that the coders with the textbook answers aren't the ones with the creative skills necessary to provide quality coding solutions. So, two years later, Inmar was looking for someone to fulfill the same contract. And we were given a second shot.

This time, we came prepared. We found out Inmar were having trouble calculating their monthly reports. That was a problem Dave knew everything about. He'd solved similar issues for companies like MedCost and Kay Chemical Company. So, before Dave went into his second interview, I called up Inmar and said to ask Dave about those projects specifically. At the same time, I told Dave to bring up those examples whether the interviewer asked or not.

When Dave walked into that interview again, he was much calmer. Because he could talk about his own projects, there was no risk of brain blank. Once he was on firmer footing,

Dave was able to ace the interview, and he walked out of the office with the contract.

Dave very quickly proved he was the best coder to interview for that job the first and the second time. It took him only one month to fix Inmar's reporting problem. In that month, he managed to reduce the amount of time those reports took to compute from forty-five days to four minutes.

Dave was always the best candidate. What changed between those two interviews wasn't his skill but how he engaged with his interviewer.

HOW SOFTWARE DEVELOPMENT INTERVIEWS WORK

It was Dave's experience at Inmar that showed me the secret to acing the coding interview: you have to hack the interview process.

Development managers play the code trivia game because their development shop usually has a particular problem, and they aren't sure how to test whether you can solve it. So, they give you the bare minimum of context and then ask complex questions to test your encyclopedic knowledge of every coding language and technique ever invented.

This would be like walking into an interview for a position as a mechanic, and because the garage has a lot of Jeeps in the back, the interviewer asks you to name every single part in a Jeep from the tailpipe to the antenna.

Not every interviewer plays the code trivia game, of course, but they may have equally unproductive and difficult tests in mind for you. Sometimes, for instance, a development manager will ask you to write out the code for an app on a whiteboard. The goal is to test whether you can code from memory. Despite the fact that with programs like Visual Studio you never have to code from memory, some development managers feel this proves who really knows how to code.

This would be an absurd interview tactic in any other profession. Imagine a writer interviewing for a job and being told they had to write the first chapter of a new book on a whiteboard with no preparation. Then, the interviewer would go up and check their spelling and grammar.

Even though the writer would work in Microsoft Word—which would catch these mistakes for them—they would still have to prove they could come up with a chapter and write it out flawlessly. Would that test actually determine which writer was best? Of course not, but this same test is carried out all the time on coders.

HOW TO HACK YOUR INTERVIEW

At root, this flaw in the interview process stems from the

development managers' poor interviewing technique. Too often, they just don't know what questions to ask to determine which candidate is the right fit for their open position. Unfortunately, this problem has existed for decades in software development, and it's unlikely to change anytime soon. The only solution is to either memorize an encyclopedia of coding terminology, or learn how to flip your interview so that you are in control of the questions.

And the best way to perform that flip is to redirect your interviewer's focus toward your portfolio.

PUT YOUR PORTFOLIO FIRST

Every time you walk into an interview, the development manager has a list of questions and tasks in mind that they want to use to test you. Your job is to make them throw that list out. As early as you can, transition the focus from the list to your portfolio. You can start by doing exactly what Dave did in his second Inmar interview: walk in and boot up your laptop. That communicates that you've got something to share, and hopefully starts the interview on the right foot.

Once your portfolio is the focus of the interview, questions will naturally revolve around the work you've already done. Instead of testing how well you remember a random fact about C#, the interviewer will ask you questions about how you developed your bug tracker or which languages you coded your blog in.

At that point, you've basically transformed your interview into an open book test. Abstract questions about coding techniques become concrete opportunities to demonstrate the skills you used to assemble your apps.

Of course, interviewers aren't necessarily eager to allow you to turn their interview upside down. Many will try to return to their trivia questions and whiteboards. They have an interview process, and they don't always like being steered away from it.

You don't have to be antagonistic in such moments. Instead, simply keep redirecting their questions to your portfolio. If they ask you a question about JavaScript, you can say, "Let me show you how I used that language in my blog."

If you get a question about a concept in C#, don't answer with a textbook definition; instead, show the interviewer that you know C# by relating the question back to your portfolio.

All you have to say is, "That's a feature in C#. I built this whole budgeting app in C#. Let me show you how I use that in my project."

Eventually, most interviewers will look at the code you're

putting in front of them. So long as you have high-quality apps in your portfolio, once they're looking at your work, you'll keep their interest for the duration of the interview.

TALK ABOUT WHAT YOU'VE BUILT

Once you've hooked the development manager with your portfolio, they'll put aside their questions and ask you about what you've built. So, you'd better be able to talk about your code.

The very first question you're likely to hear from an interviewer reviewing your portfolio is, "How did you build this?" Far too many coders get to this moment and then fail to provide quality answers. They're not prepared to talk about their programs on a technical level.

That's why, at Coder Foundry, we train our students to talk like coders. We teach them how to talk about their portfolio on a detailed level. They don't have to be able to answer all the code trivia questions, but they do have to be able to talk about the code they've already used.

For example, let's imagine your interviewer starts out asking you to define MVC. You don't need a textbook definition, but you do need a basic answer here. So, you can say, "MVC stands for Model-View-Controller. It is a design pattern used for building websites." Then, you can pivot toward your portfolio. "As you can see, I used MVC to construct my bug tracker. Just take a look at the MVC design pattern here where I wrote the code to pull information from the database. Do you mind if I take a few minutes to walk you through the code?"

By demonstrating a knowledge of the languages and techniques within your programs, you'll pass every possible test even the most skeptical development manager could have devised. Instead of playing their games, you'll have shown them you have the exact skills and knowledge they need.

RECOVER FROM BRAIN BLANK

Dave suffered the most notorious brain blank I've ever seen in an interview, but he's got plenty of company amongst the failed code trivia players out there. Brain blank happens all the time in software development interviews. It can happen to the most skilled and experienced coders, and it can certainly happen to you.

If you follow the advice in this chapter, however, you'll be able to protect yourself against the scourge of interview brain blank. In those moments when your memory fails, you can rely on your secret weapon: your portfolio. When you can't remember exactly how a certain C# feature works, simply walk the interviewer through your C# code until you come across the feature you need. If you've built an impressive portfolio and know how to talk about it, the interviewer will be so dazzled, they won't care much about the answer to their trivia question anyway.

This hack works almost every time. It's a large part of the reason we place almost all of our students in fast-advancing positions right out of boot camp.

However, if you want to use this technique, you have to get an interview first. Unlike other professions, you can't do that on your own. In this business, there's a step between the boot camp and the interview: the recruiter.

CHAPTER 7

STEP 5: WORK WITH A RECRUITER

In chapter 1, we met fifty-two-year-old Tom Harrison, who feared he was too old to change careers. It turned out that he wasn't too old, but it was tricky getting him that first job. Like every other software developer, he needed a recruiter in order to secure interviews, but the top recruiting agencies weren't interested in him at first.

A recruiter is the middleman between the software developer and the companies that need their services. They know where the jobs are and who fits the job description. Then, they use that information to pair quality coders with excellent positions.

The system works very well, but there is a flaw. Recruiters often have an ideal software developer they want to work with in mind. They want someone young, already experienced, and eager to take the biggest paycheck they can get. Recruiters will work with coders who don't match that description, but the coder has to convince the recruiter to do so.

Because he didn't match the profile, Tom had to use his portfolio and interview skills to get himself into the profession. He called a recruiter at Tech Systems, one of the biggest national recruiting firms for software developers.

"Hello," he said, "My name is Tom. I've just completed my coding boot camp, and I was wondering if you could help me find a job."

As expected, the recruiter initially wasn't interested.

"I don't work with inexperienced people," he said. "I want senior-level people with three to five years of experience."

Again, that's not an unusual answer, so Tom wasn't put off. He'd been coached for situations just like this. He knew just how to react, and he followed the script perfectly.

"That's great," Tom said. "I get it. But can you tell me what types of positions you're looking to fill these days?"

"Coders who know the .NET stack."

"Perfect, that's exactly what I learned in boot camp."

"Okay, but I'm looking for someone with experience I can sell to organizations."

"I understand," Tom said. "But would you take the time to look at my portfolio and give me some tips that I could use to break into the industry?"

The recruiter paused. "You have a portfolio?"

Now Tom had him. He explained the programs in his portfolio and gave a detailed account of how he built them using every tool in the .NET stack.

As soon as Tom had finished explaining, the recruiter's whole attitude changed. He asked Tom to send the portfolio over right away. Within two weeks, Tom had been placed in his first position. He's never looked back since.

LIKE IT OR NOT, YOU NEED A RECRUITER

Here's a truth some coders out there may not want to hear: in this industry, you have to work with recruiters. It doesn't matter if you like the system or not. It doesn't matter if you're really independent and want to do it all on your own. If you want to work as a software developer, you need a recruiter.

There are many reasons to use a recruiter, but perhaps the most important is that without a recruiter, you may never even get in the door for an interview. All the time, I hear sad stories about coders who have applied to thousands of positions and never even received a reply back. They didn't hear anything because, by the time they applied, those positions had already been filled.

Let's say you're scouring the job boards online for a position as an entry-level web developer. This is the most in-demand job in the country, so every day, when you check Monster, Indeed, and Dice (which specializes in tech jobs), there are new postings. You apply within minutes to every one of those jobs. Yet, somehow, you never hear back.

The problem isn't you, it's the system. Every time you respond to one of those posts, your resume gets dumped into a pile with a thousand other applications. A resume scanner—a program, not a person—then checks it to see if you qualify.

If your resume is not formatted correctly, or a typo misstates your education or experience, your application is thrown out—even if you're perfect for the position. Should your resume pass the scanner test, it goes to the bottom of a very large pile of applications, where the HR manager may never see it.

All that is beside the point, though. Even if your application somehow landed at the very top of the pile, you're still unlikely to ever get an interview. Tech Systems and every other recruiter out there have already filled that position. Part of a recruitment agency's job is to call HR departments every week to find out about upcoming positions. Before that job ever gets posted, a recruiter has already recommended ten well-qualified candidates to fill it.

So, even if you are the first person to see a posting and apply to it, even if you pass the resume scanner and have ideal qualifications, you're still at a huge disadvantage. By the time your application gets into the hands of the development manager who would hire you, they've already interviewed candidates from resumes a recruiter sent over last week.

That's what you're up against when you are checking the online job boards 24/7. That's your best-case scenario, and in fact, that's only the beginning of your difficulties. About half of all software development jobs are never even posted. This is what we call the "hidden job market." These jobs are *only* sourced to recruiters. You can't even apply for them unless you're already working with a recruitment agency.

Why is the industry like this? Because recruiters provide businesses with a valuable service. They take care of the vetting process by testing new prospects. They interview software developers, review their portfolios, give them coding tests to check their knowledge, and double-check their resumes. So, when a company uses a recruitment agency, they can be sure that the only potential hires who walk into the development manager's office have already proven they are capable of doing the job. This saves companies time and ensures they don't accidentally hire someone underqualified for the work.

If you ran an HR department, would you rather interview eight hand-picked candidates from a respectable recruitment agency or spend weeks going through thousands of applications to see if you could find those eight people yourself?

The recruiter is simply the easier, more efficient choice.

They're also the better choice for you, because recruiters don't just work for these companies; they work for you. Part of their job is to sell you to your potential employer. They want you to look good because they want you to get that job. That's how they get paid.

So, if you flub a code trivia question in an interview, your recruiter can get on the phone and say you were just nervous. Or, you're naturally quiet and reserved. They can redirect the development manager to review your test results and your portfolio again.

A call like that from a respected recruiter can turn your prospects around. Whereas, if you just come into an interview unrepresented and suffer a brain blank moment, there's no one to advocate for you afterward. It's just you.

LIKE A PRO ATHLETE'S AGENT

If you're still hesitant about being represented by a recruiter, think of yourself like a pro athlete. How often do you hear about a pro football player representing themselves without an agent? It happens, but it's extremely rare.

If you want to be a running back on my favorite team, the Las Vegas Raiders, you don't send over an application; you send your agent to talk directly to the coaches and the general manager. Your agent will be the one who markets you to the team.

His job is to build up all your successes and bury all your weaknesses. If you had a great combine, he'll push your stats. If not, he'll talk about your college career. Whatever he has to say to make you sound fantastic, he'll say it.

His aim is to make sure the Raiders don't just want you—they need you. Then, he makes sure your salary reflects that need. If he can't get a deal with the Raiders, he uses his connections to sell you to every other team in the league.

That's exactly what representation does for you in the software development field. Like the NFL, this is one of the few industries where you have an advocate on your side before you take your first job.

And a recruiter's services are even more valuable than an NFL agent's. The NFL only has thirty-two teams. Each team has the same number of players and very publicly needs certain positions filled. In IT, all the companies are unknown. You have no idea where those hidden jobs are. There could be a fantastic opportunity at company X right down the street, but only your recruiter knows about it.

THERE IS NO DOWNSIDE

Despite all the positive work recruiters do in this business, many coders are still biased against them. In particular, there's a lot of negativity about recruiters on social media—which is yet another reason to avoid getting your coding career advice from Twitter.

Much of the criticism about recruiters stems from the fact that they get a percentage of your first year's annual salary. But before you turn away and swear off recruiters forever, you should know that *you don't pay recruiters a cent*. The company that hires you pays that percentage *on top of the wages they give you.*

So don't listen to those negative reviews online. Yes, the

recruiter who finds you a job will get compensated for their work, as they should. They aren't running a charity. However, because they get a percentage of what your employer pays you, they are incentivized to always negotiate the highest possible wages for you.

There's no downside to using a recruiter. They have access to all the jobs and the ability to put your name at the top of the application pile. They sell you better than you can sell yourself, and they make sure you always get paid top dollar. For all that assistance, you never give them a penny of your earnings. This is a win-win-win system for coders.

RECRUITER OPTIONS

You need a recruiter to navigate the software development industry, and luckily, you can find recruitment agencies all across the country. Two of the biggest—with national reach—are the previously mentioned Tech Systems and Robert Half. Additionally, there are many local and regional firms.

Your first recruitment option, then, is to apply to either the national or local firms the same way Tom did: call up a recruiter and sell yourself using the strategies we covered in the last two chapters.

If you are reticent about cold calling recruiters, Coder

Foundry offers another way to get the same advantages in the industry.

We are unique in providing our students not just a boot camp but also the services of a national recruitment agency once they graduate. We have the same reach as the big firms, and we will advocate for you across your career with the same determination.

RECRUITERS WILL TAKE YOUR CALL

If you don't come to Coder Foundry, though, you will have to make those cold calls. But don't worry—recruiters will pick up the phone. They will hear you out, and they will review your portfolio.

They have to. Part of their job is to talk to a certain number of software developers every week. They have to show their bosses that they took fifty calls and reviewed fifty portfolios and resumes between Monday and Friday. That's a lot of calls, so they will actually be grateful when you make their lives easier by reaching out yourself.

In order to remain successful, a recruitment agency must constantly bring in new talent to fill all their open positions. Since more and more companies need software developers, recruitment firms have to keep looking for new people to add to their database.

With a solid knowledge of the .NET stack, a strong portfolio, and the ability to talk about your code, you'll be an excellent candidate to put into any recruiter databases. Then, your recruiter can pull your profile out as soon as the next open position pops up.

BUILD A RECRUITER RELATIONSHIP THAT LASTS

Your relationship with your recruiter doesn't have to end as soon as you land that first job. Instead, that relationship should continue throughout your career. If you build a strong, professional relationship with your recruiter, and you're good at your job, you'll find that your recruiter can place unbelievable opportunities in front of you.

But they'll only do that if you make this relationship a priority.

In the software development industry, it's not your job to find your next high-paying job; it's your recruiter's job. They are always looking for your next position, one that will pay better and offer more rewards. If you're making $50,000 a year, it's your recruiter's job to find a company that will bump your pay up to $75,000. If you want to move out to Silicon Valley, it's your recruiter's job to find that position. In short, your recruiter is your individual economic mobilizer.

At times, they can be even more than that. My first recruiter,

David Bryant, has become a lifelong friend. I met David in the early '90s when he helped place me in one of my first jobs. When I decided to move on from that position, he located multiple opportunities for me. Then, when I founded Core Techs, he helped me bring talent into my new organization.

Whatever direction I wanted my career to go, David was there to help me fulfill my goals.

You don't have to develop a friendship with your recruiter like I did with David. They won't expect an invitation to Thanksgiving or a Christmas card every year. However, they will expect you to answer their calls and respond to their emails.

When they reach out to you, always reply promptly. If you don't, they aren't going to send you angry texts, but they will drop you down their priority list. Then, when your dream job arrives, they'll hand it to a software developer who stayed in contact.

If you remain professional and engaged, your recruiter will always be there for your next career advancement. That's a powerful ally you don't want to walk away from.

FIVE STEPS TO THAT FIRST JOB

You can draw a clear line between those coders who followed the five steps I've laid out in this book and those who didn't. The ones who followed them have excellent jobs in the industry. Those who didn't, sit around online complaining about the lack of positions in software development.

Far too often, I meet someone at a conference—or come across them on LinkedIn or Twitter—who says the industry is rigged and no one is hiring. When I inquire into their background, I always discover the same things. They always did one of the following:

1. They learned the wrong stack
2. They didn't learn practical coding skills in school
3. They have a weak (or no) portfolio
4. They struggle in interviews
5. They don't have a recruiter

In software development, getting that first job is really as simple as working your way through these five steps, one after the other, just as you've learned in part II of this book.

Once you complete the steps, you're set up for success—at least initially. There's still far more you need to know before you can go from having great prospects in this field to a real, economically mobilized software development career.

I want to get you into that first position, but I also want you to reach the highest levels of your ambition. For that, we have to move beyond these five steps and look at how you walk into that first job and advance to your future success.

PART III

THE FIRST JOB AND FUTURE SUCCESS

———

CHAPTER 8

TAKE THE FIRST JOB

Two of my best coders, Jacob and John, graduated from the same boot camp. Both were excellent students with fantastic portfolios, and both had bright futures ahead of them.

Jacob was nineteen at that time. He was passionate, smart, motivated, and extremely excited to start his career. As soon as he graduated, a recruiter got him an interview for a marketing company in Georgia. Jacob drove five hours down to Atlanta from Coder Foundry.

I expected a call after the interview telling me he had the job. But there was no call. When he got back to North Carolina, he walked into my office and told me that he had been offered the job right there. His career was made, but he wasn't celebrating. He looked unhappy. I asked him what was wrong.

"I can't take the job," he told me.

This was 2016, and in one of the marketing company's office windows, there was a big campaign sign for a candidate Jacob had very strong negative feelings about. It was clear to him from his interview that the owner's politics were very different from his own.

Everything else about the job was perfect: a great location, interesting work, and a good starting salary at $47,000 a year. But Jacob didn't want to take it. He said it would compromise his beliefs.

I told him to really think about what he was doing. With software development, you have to take that first job offer. There are lots of opportunities down the road, but there's no guarantee after you get that first offer that another will come along.

I told him that this company wasn't going to do anything immoral or illegal. He never had to discuss politics in the office. In fact, he could put bumper stickers on his car for his preferred candidate and spend almost all his time in a cubicle with his headphones on. If he really felt so strongly, he could donate part of his salary to a campaign he supported.

"In a year," I said, "you can start looking for something else. Once you've got that little bit of experience, you can go work for someone who shares your beliefs. By all means, vote and volunteer for what you believe in, but don't pass on this opportunity."

He didn't listen. He walked away from the job, and to this day, he's not in the software development field.

John was less political but no less firm in what he wanted from his career. The most important thing to him was to find a position in North Carolina. That's where his family was, where his home was, and where he wanted to live out his days. John's first job offer, though, wasn't in North Carolina; it was at a health care organization in rural West Virginia.

To put it mildly, he didn't love the relocation idea, but he took the job anyway. He moved out to West Virginia. Only six months later, he moved back to his hometown of Greensboro for a job that paid him $15,000 more.

Nothing separated Jacob and John in classroom performance. In theory, they both had the same great career prospects. The only difference was one took that first job when it came for him, and the other didn't. Now, one is making great money and moving up in the software development world—living right where he wants to live—and the other is trying to get by on other skills.

NO ONE WILL HAND YOU THE PERFECT JOB RIGHT AWAY

I don't fault Jacob for having strong political beliefs. I think it's wonderful when people feel passionately about the world. I loved Jacob's idealism and convictions, and that idealism is not a rare thing in software development.

Most of us enter this line of work with an ideal in mind. It isn't always about politics. Often, like John, it's an ideal life we want to live. We want to live in a certain place, do a certain kind of work, and achieve certain goals. Maybe we've always wanted to live in New York City, or maybe we've always wanted to help promote great scientific research. Regardless, those ideals are firmly held.

With software development, you can realize any dream or ideal, but you may not be able to do so with your first job. You can get there, but you have to start at the first place someone is willing to give you a shot.

The perfect jobs out there don't want to take a chance on someone just out of school or boot camp. They wait for smaller companies to take those chances. Then, once a coder has proven they can handle the work, the dream jobs jump in and make an offer.

When you can combine a strong portfolio with a year of experience, you can become truly unstoppable in this industry. But that experience is crucial. To move up, you have to start by getting a company—almost any company—to pay you to write code.

Of course, I would never recommend that someone take a job where they would be seriously underpaid or mistreated. If you walk into an interview and the roof is falling in or unreasonable demands are put on you, by all means, walk away. Otherwise, you have to take the first opportunity when it presents itself. That's the only way to get the perfect job down the road.

WHAT HAPPENS WHEN YOU PASS ON THE FIRST JOB

While the five steps we covered above will place you on the path for any kind of success in software development, those steps only work when you take advantage of the opportunities in front of you.

When you pass on the first job, you can become yesterday's news faster than you might think. In boot camp, you're in an immersive environment, learning the whole .NET stack in twelve weeks. To keep that knowledge fresh, you need a job as soon as possible. The longer you take to find that perfect first position, the faster your coding skills erode. It's hard to keep practicing and maintaining your motivation without a paycheck. With bills to pay, you may settle for an unrelated position while you wait for the right opportunity.

Once you're out of boot camp, you won't be surrounded by a community that can support you, and your mentors will be focused on new students. At the same time, recruiters quickly lose patience with new software developers who refuse good jobs. It doesn't make any financial sense for them to keep finding positions for someone who won't take them. So, you may find yourself working on your coding career on your own, trying to balance coding practice, a day job, applications, and interviews.

Within a few months, the newest students will have graduated with their portfolios and skills fresh from boot camp. They'll be eager and motivated. They won't have any rust on their abilities or their understanding. Suddenly, you're competing against a whole new set of coders, most of whom will take those jobs you didn't want.

Very quickly, you can see all your opportunities dry up. After a year, your boot camp degree will look old, and you'll have no work experience to bolster your resume. Your skills will be out-of-date, and you'll have few if any connections to leverage into another chance at an interview.

That's how talented coders like Jacob go from the best prospects to permanent failures to launch.

YOUR FIRST JOB ISN'T YOUR LAST JOB

Even if you hate your first job, I promise you, it won't be the end of the world. To begin with, your recruiter is not going to let you interview for a bad job. It makes no sense for them to put you in a position with an awful wage where you want to quit immediately. That would harm their reputation and their bottom line.

So, when you do get placed in a job, it's not going to be that bad. You're not going to be breaking rocks outdoors in subzero temperatures. You're going to be sitting in a comfortable chair, in a warm office, writing interesting code for a good wage.

I'm not saying that first job will be amazing. There might be a bad culture in the office. You might have a boss who acts like a jerk. You may have to work in a location that isn't ideal or in an industry you have no interest in.

For many people, such a position would represent the end of their opportunities, but that isn't the case in software development. You don't have to stay in a job you don't like. If you put your head down, put your headphones on, and write code for a year, you can move on to something far better.

With the demand for software developers in today's economy, that little bit of experience earns you entry into some of the most desirable positions in the country. Every company is begging for experienced, proven coders. That's why you always hear about IT talent getting the VIP treatment with ping-pong tables, free soda, and playgrounds in the workplace.

Once you've finished your time on that first job, the economic mobilizers really kick in. You can contact your recruiter, tell them to find you a job to your specifications, pack your bags, and settle into a job more to your liking.

ALL IT TAKES IS ONE YEAR

Your career prospects really open up once you have a year of experience. One of my former students, Sean, came to Coder Foundry after deciding he wanted to transition out of his first career as a detective. His goal was to code for a huge corporation. For some reason, he loved the idea of a big corporate structure he could slowly climb.

His first job offer wasn't at a massive multinational, though. It was at a small company called AvaCare. When he got the offer, he was unsure if he wanted to take it. I told him exactly what I told Jacob, John, and now you: "This is not your first job, and it will not be your last. Take this position. The next one will be at a big company."

He took the job, made $65,000 during that first year, and then moved on to a big corporation where he now makes the in-car software for police officers and first responders. He's got the job of his dreams, but it was only possible because he took the first job that came along.

Another student, Luke Sanders, graduated three years ago. He's had three jobs in that time: the first job paid him a modest starting wage, the second paid a little better, and his third job now pays him $110,00 a year.

He took the first offer he got, transferred into the second job as soon as he could, and now makes more than he ever dreamed he would.

Learning to code can change your life incredibly quickly, but it still takes a little bit of patience up front. The companies with the best pay and most interesting jobs will hire you, but only after you gain some experience.

Once you've proven you can handle a professional software development job, you can start setting your sights as high as you want. In fact, you'll be one of the top candidates for most of the jobs you apply for. Instead of taking whatever you can get, you'll be deciding among three, four, or more offers at a time.

Your job prospects increase the more your experience grows. Once you have three to five years of experience, you can start working with your recruiter to target specific companies, locations, and wages. You can call up your recruiter and say you want to work for Google, and at that point, they can make it happen. Or, if you're passionate about your politics like Jacob, you can tell your recruiter you want to work for a particular nonprofit or even a political campaign, and that can happen, too.

This is exactly why Tom Harrison's recruiter in the last chapter wanted to work with coders with three to five years of experience. Placing experienced coders is easy money for recruiters. However much they want to make, wherever they want to work, it's as easy as calling up the company and handing over the portfolio and resume.

In five years, you can reach the senior level of software development.

Alternatively, with that level of experience, you'll be perfectly placed to start your own business. You'll have the income, the know-how, and the connections to immediately become a player in your chosen software development niche.

That's how I came to run my own companies. And that's how you could end up after training at coding boot camp and getting three to five years of experience on the job. You can become the next Mark Zuckerberg.

Once you have the experience, the only potential hurdle between you and any kind of success is your dedication to mastering all the necessary coding languages to complete the projects that interest you.

That's why you have to use your time well and specialize in your true coding interests while you're building that experience.

{ CHAPTER 9 }

SPECIALIZE IN YOUR FREE TIME

Many of the students who come to Coder Foundry have a particular direction for their career in mind. They want to be a game developer, work with AI, or code the next iPhone. We have to address that dream early on in their education because it can lead to them passing on the first web development job that comes their way.

Apple, Microsoft, and Activision don't come calling as soon as you've learned the .NET stack. They want you to have experience, and they also need you to know more than a coding boot camp can teach you.

There are hundreds of coding languages out there. Web development jobs most often focus on the .NET stack. However, apps, games, AI, and phone operating systems usually work with different languages. If you want to compete for the most in-demand jobs in the coding field, you have to walk in the door knowing all the necessary coding languages and with a portfolio that demonstrates that knowledge.

As I've said, in order to break into the industry, you need to put aside the "cool" languages and focus on the .NET stack because it can deliver you that crucial first job. Once you have your first job and enough experience, then it's time to take those cool languages off the shelf and start practicing.

SPECIALIZE ON THE WEEKENDS

Almost everyone wants the same cool coding jobs you've always dreamed of achieving. Positions in fields like game development or at world-famous companies draw applications from coders around the world. If you choose to throw your own application onto that pile right out of boot camp, most of your competitors for the job will have far more experience and a vastly more impressive portfolio. They'll walk into the interview knowing a dozen coding languages all relevant to the work ahead. In other words, they'll be ahead of you on every important metric.

To get those jobs, you'll have to combine experience with the specialized skills those jobs require. But how can you do that when your first job—that you have to take—is a .NET stack web development position?

The shortest route from web development to any software development specialization is coding on the weekends. Work your nine-to-five through the week, get your paycheck, and then, in your free time, have fun with expanding your coding skills. You can start by developing games using C#. By the time you finish your first year, you can add your best game to your portfolio. Then, you have something to show those game development companies.

At the same time, you can start learning new coding languages to expand your capabilities. There are plenty of free online resources to help guide your continuing coding education. While these resources are a poor way to learn to code initially, they are much more effective once you understand the basics behind software development.

So, on Saturdays, you can sit down, watch an online coding tutorial, take an online test, and once you have the new language down, start developing a game or app in your new language.

As soon as you master one language, switch over to another. Every time you complete a language, build another project and put it in your portfolio. Within a year, you'll no longer look like a coding novice; you will look like an expert who can handle any software development task.

At that point, you can call your recruiter, tell them the new languages you have mastered and the projects you have in your portfolio, and explain the type of position you want them to find for you.

Your recruiter will do the rest. They'll find a company that meets your demands with an opening for a coder with your skills. Then, they'll sell you and your skills to that company. All you have to do is follow the exact same interview advice we covered above, and you should be able to walk into the job of your dreams.

LEARNING THE ROMANCE LANGUAGES

If you're worried you won't be able to pick up new coding languages because they will be too complex to learn on your own, I have great news for you. Most coding languages use the same basic logic. They often share a lot of the same syntax—which are the rules and symbols that govern communication with the computer. Once you have mastered the underlying logic and those shared syntaxes, learning new languages is far easier than learning your first one.

In this way, coding languages are very similar to Romance languages like Spanish, Portuguese, and Italian. Learning your first Romance language is challenging, but once you know it, it's far easier to pick up the next one because the languages share a lot of vocabulary and grammar. Even French, which is more distinct, has huge amounts of overlap. Learning the first language takes serious effort, time, and energy. But learning each related language gets increasingly easier.

Many coding languages are what we call "C-style languages," which means that they all look and behave very similarly. Therefore, since you'll already know how to do code in C# after boot camp, all you'll have to do is expand upon that knowledge with related languages like C++ or Java.

Code in Javascript

```
for (i = 0; i < 5; i++) {
  text += "The JS number is " + i + "<br>";
}
```

Code in C#

```
for (int i = 0; i < 5; i ++) {
  text += "The C# number is " + i + "<br>";
}
```

In some places, coding languages can be almost identical.

Essentially, you'll mostly be learning new techniques instead of whole new languages, and the more you learn, the easier it gets to incorporate and integrate these new coding languages into what you already know.

PETER JACKSON DIDN'T START WITH *LORD OF THE RINGS*

The process you have to follow to get the job of your dreams can sound frustratingly indirect. You want to work as a game developer, and instead you have to spend a year doing web development and committing your weekends to learning new languages and building new projects. Why can't you just learn the cool languages first and jump straight into that cool career?

As we've seen already, trying to cut the process short often cuts you out of the process entirely. This is just how the software development industry works, and it isn't unique. In fact, most creative fields require that you work your way up in tangentially related fields, developing diverse skills and proving your creative flexibility.

Rarely, if ever, will you hear about Hollywood handing a novice director control over a major motion picture. Peter Jackson is world-famous these days for directing his passion projects, the *Lord of the Rings* and *Hobbit* trilogies. At this point, he's so well-regarded that he can probably get any project he wants financed (as evidenced by the recent flop *Mortal Engines* that he wrote and produced).

He grew up a huge Tolkien fan, but he didn't just walk into the offices at New Line Cinema and demand to direct those movies. Instead, he started his career directing low-budget horror flicks on the weekends. He was only handed the responsibility for his dream project once he'd learned all the necessary skills and demonstrated real promise on a much smaller scale.

In the same way, you have to prove that you are capable of the cool, romantic coding job you want. Once you've shown you can handle the responsibilities of web development, along with the motivation to learn new coding languages outside the office and the skills to create projects on your own, then your ideal employer will take notice.

If you can take the opportunities that come your way and learn how to communicate effectively in your job, then you can move up to that ultimate Peter Jackson-level position where you can truly call the shots and reap all the benefits.

CHAPTER 10

COMMUNICATION ROCKS

There are two kinds of software developers in every workplace: grumpy coders and rock star coders. These coders aren't separated by skill. They may both write code exceptionally well, and when given a project, they may deliver similar results at the same pace.

The difference is their attitude toward their work and their colleagues. A grumpy coder makes life difficult for everyone at his job. He doesn't share information. If he is asked how long he needs on a project, he responds moodily, "It'll be done when it's done."

He won't share how his code works or how he arrived at certain choices. If he is ever questioned, he becomes defensive. At the same time, he overestimates his value to his company. He feels that he is the most important person in the office simply because he knows how to code.

He undervalues everyone else's experience, knowledge, and expertise.

A rock star coder, on the other hand, loves to share. She's happy to explain how everything works. She's willing to work on a timeline and work through issues in the code whenever they arise. She empathizes with her coworkers and recognizes that everyone has value. She sees sharing her expertise as part of her value to the company.

Over time, these two coders are going to experience markedly different careers. The grumpy coder is going to stay right where he is, if he's lucky. The best he can hope for is that his skills are good enough that he is too valuable to fire.

Even if he's specializing on the weekends, he's never going to land that dream job. Or, if he does, he won't be able to keep it. He's also not going to move up at his current company. His prospects are limited to the dimensions of his cubicle.

Whereas, for the rock star coder, there really are no barriers. She can become her company's CEO or move into a cool, specialized software development field like gaming. Everyone will want to help her rise to her ambitions.

For the rock star coder, the sky is the limit.

THE DIFFERENCE A ROCK STAR PERSONALITY MAKES

When I say the sky's the limit for rock stars in this field, I mean it. Once you have the experience and specialized knowledge, being a rock star is what truly elevates you above the competition to reach unbelievable success.

Any coder is capable of becoming a rock star coder. Being a rock star in this field isn't about being cool; it's about being passionate, approachable, and communicative. You can see all of those qualities in the two biggest rock stars in the history of software development: Bill Gates and Steve Jobs.

No one has ever accused Bill Gates of being cool, but his enthusiasm is contagious. He set out to change the world, and in his days as CEO of Microsoft, that aspiration was palpable whenever he was on screen. People bought Windows and Office partly because Gates could explain to them why those products were so much better than the alternatives on the market.

Steve Jobs was even more of a rock star. When he stood on stage in his iconic black turtleneck and held up a device, his rock star passion made the whole world pay attention. He demystified his complex technology so everyone could understand it and understand they wanted it. That's part of the reason Apple is valued at a trillion dollars today.

Clearly, rock star coders can make unbelievable amounts

of money for their companies—and for themselves. More than that, a rock star personality can change the world.

THE PRICE OF GRUMPINESS

Rock stars change the world; grumpy coders get left behind. I want to focus on the cautionary side of this tale because, unfortunately, there are far too many grumpy coders out there. I've worked with many of them, and I've seen them get passed over for every advancement.

How much grumpiness will harm your career depends on just how grumpy you are. There's a spectrum of grumpy coder behavior that runs from being unnecessarily standoffish to refusing to finish projects on a timeline. The reserved grumpy coder can at least hope to keep their position; those on the far end of that spectrum may want to look for a different career.

Regardless, by indulging in your grumpiness, you are only harming yourself and putting limitations on your prospects.

In contrast, a rock star coder can actually be less knowledgeable, less experienced, and less adept at coding, and they will still be more likely to advance. The fact that they invite others into the process, deliver on time, and allow others to understand what they're doing can make up for significant deficits in their skills.

Because people like working with a rock star, she'll move up, while the grumpy coder is still sitting in his cubicle, thinking that he's better than the rest of the world.

REAL TALK FOR GRUMPY CODERS

Once you've mastered all the tips we've covered in this book, improving your social skills is the best way to advance your career. This doesn't mean you have to become a charismatic socialite capable of climbing onstage to sell the next iPhone in a black turtleneck. The social skills you need in this profession are as simple as learning how to hold a friendly conversation with colleagues and showing respect and empathy to others in your office.

I don't want to underestimate how difficult even these skills are for some coders. Many "failure to launch" individuals have struggled to enter into a career precisely because they are intensely introverted. They become anxious trying to make small talk, and they aren't naturally adept at reading social cues. Luckily, being a rock star coder doesn't require constant socializing; it requires openness, straightforwardness, and a willingness to share what you know.

Sharing is the real dividing line between grumpy and rock star coders. Grumpy coders suspect that if they share what they know, they'll become less valuable. They see their knowledge as a finite resource. They believe that once it's shared, others will be able to do the work, and they will have communicated their way out of a job.

That's simply not the way business works. No one wants to replace the software developers on their team. The job is going to remain valuable, even if others have a basic understanding of how coding works. In fact, sharing that information increases a coder's value because everyone can better appreciate how important their work really is.

Rock star coders recognize this fact, and they use it to their advantage. By sharing their knowledge and their passion for their work, they demonstrate just how critical they are to their organization, and how useful they could be in even more powerful positions.

ROCK STARS ARE COMMUNICATORS

Becoming a rock star coder doesn't have to be particularly taxing, even for those who are naturally introverted. Some social skills have nothing to do with interacting with others. One very basic rock star trait you can develop is an appreciation for the abilities and knowledge of colleagues. Many grumpy coders treat those who don't understand code like they're stupid. They have an attitude that suggests, "This is *far* too complicated for *you* to understand."

A rock star coder recognizes that the people they work with

may not understand how to code, but that doesn't make them less intelligent, or less important to the organization. They simply have a different kind of knowledge, a different set of skills, and a different category of experience. Remember that software development is only one component of the overall strategy that makes a company profitable. A software developer is a part of a team in which every person is working toward the same goals. The more the team can work together, the better everyone achieves their aims.

So, when you interact with others, you don't have to race to the water cooler and prove you're funny. Instead, build relationships based upon respect and empathy. On some level, your communication should all come down to what Mark Cuban calls the new superpower: being nice.

Being nice, you can engage in conversations with a desire to be helpful. When your boss needs to know the timeline of your project, give her your best estimate. Work on developing a way to make your work seem less mysterious to those without a technology background. You don't have to teach anyone to code, but you can give them a basic idea of how an app works and why it takes so long to build.

Channel your internal Steve Jobs and find the language to explain the highly technical in a nontechnical way. Remove the technical terms and acronyms from your explanations and compare your work to something others find more relatable. Simplifying your very complex work doesn't make you look less clever; it makes your work more impressive because people have a better idea what you're doing. Once people have that understanding, they can better appreciate your creativity and skill. With a little rock star communication, your code will no longer look like nonsense to your colleagues; it will look like the complex, creative project it is.

Essentially, communication is the key to becoming a rock star. When you explain to your team how brilliant, cool, and complex your work is, they aren't going to replace you. They're going to give you more responsibility, more authority, and more income, because they want more of what you're offering.

At that point, you become truly irreplaceable, and your skills make you an ideal candidate for even greater success—either at that business or elsewhere.

COMMIT TO THE WORK AND THE TEAM, AND YOU'LL GO FAR

Software developers aren't always responsible for the grumpiness in an office. There are grumpy salespeople, grumpy marketers, grumpy managers, and grumpy personal assistants. However, your job as a coder is to be the rock star in that grumpy office. Even if you work with an overbearing salesperson who only cares about his P&L

reports, your job is to build a bridge in that relationship. Practice empathy and communicate your ideas clearly, and leave the grumpiness to others.

Working with such people can be difficult, but new opportunities will arise that free you of that difficulty, so long as you remain a rock star. When you find yourself in a grumpy office, focus on the joy you feel in coding. Relish the creativity in your work. Continue to learn new coding languages, build interesting projects, and share your passion with everyone interested.

If you remain committed to coding and your rock star attitude, nothing—not even the grumpiest software development shop in the world—can stop you from reaching the pinnacle of your economic mobilizers and career potential.

CHAPTER 11

STAY COMMITTED

A few years ago, when software development boot camps were still fairly new, I was representing Coder Foundry at a conference for new coders called CodeNewbie. I was on stage with a number of successful coders from all walks of life, including a representative from Microsoft.

During a round of questions from the audience, a man stood up and said, "I heard Microsoft won't hire people from boot camps. Why is that? Does that mean I shouldn't go to a boot camp?"

Everyone on stage looked down and shuffled their feet. The representative of Microsoft cleared his throat to speak, but I jumped in first.

"I don't know what Microsoft's policy is. I don't care, and I don't think you should care either. If Microsoft won't hire you just because you received a great education from a boot camp, you don't want to work there.

"And you don't have to. If you learn how to code, you can work anywhere—from the warehouse down the street to a Fortune 500 company in downtown Los Angeles. You can make $100,000 a year with the best 401(k) in town, or you can found the next Microsoft and become a tech billionaire. Don't base your choices off the preferences of a couple companies. There are thousands of employers who would have you the day you graduate from a boot camp. What I'm trying to tell you is this: if you stay committed to your craft you will always have a job, and you'll always have a job you can love!"

The Microsoft representative was flummoxed, but the audience loved it. And for good reason: everything I said was true.

In the vast field of software development, one or two companies may say they don't want to work with you, but you can still work for everyone from a mom-and-pop startup to a huge Silicon Valley tech company. Or, you can found your own company and make your own fortune.

Mark Zuckerberg didn't have to go work at Microsoft. He created Facebook on his own, and now he's among the richest and most influential people in the world.

Within the next five years, the next Zuckerberg will deliver the next big advancement in technology. In fact, we've already seen a number of software developers in that mold. In the last decade, entrepreneurial software developers have invented world-changing apps like Instagram, WhatsApp, Snapchat, Uber, Lyft, and DoorDash, just to name a few.

Your company could be the next one on that list. Or, you could work for that company. It's all open to you.

And these days, if you want, you can even work for Microsoft. Since that conference, Microsoft has realized that they were missing out by refusing to recruit coders with nontraditional educational backgrounds. Coder Foundry has even sent some of our students to Microsoft, where they are still working.

Whether you want to conquer the world or settle into a steady, high-paying position with amazing job security, it's all there for you. All you have to do is stay committed, and you'll always have fantastic prospects.

YOUR CAREER CAN BE WHATEVER YOU WANT IT TO BE

James Newby is one of four Coder Foundry graduates we have placed at Microsoft. When he came to us, he was worried about prejudice in the tech world. As a black man, he thought no one would hire him as a coder. He's now a software developer at one of the most prestigious and profitable tech companies in the world.

Lauren Stewart was offered a similar position at Microsoft. After she finished at Coder Foundry, she took a Microsoft internship out in Redmond, Washington. The job was interesting, but after her internship ended, she decided she wanted to go home. So, she turned down a full-time position with the company and returned to North Carolina to pursue a successful software development career in the state she loves.

Both James and Lauren are extremely happy with their positions. They're both working hard and setting themselves up for a lifetime of success in this field. That's the most amazing part about this career. Once you're in this industry, just staying committed and working hard is enough to achieve your dreams, whether those dreams involve a massive tech company outside Seattle or a small-town business in North Carolina.

So long as you put the work in, it's entirely your choice.

ALWAYS KEEP LEARNING

The key to continuous success in this industry is to keep learning. The skills you learn at boot camp will give you a great start, but part of your job going forward is to keep developing as a coder.

Even after you successfully specialize in your preferred software development field, you should continue to learn new languages and language stacks. Additionally, you should find out how the latest software works and keep up with new techniques.

Ultimately, you should view coding as your craft. In the same way a painter never stops exploring the potential in their art, you should constantly push yourself to achieve something more from your code. Painters are always driving themselves to try something new. They work with new paints, canvases, and tools. Even when they achieve success in a certain style, they still branch out and explore other potential avenues to accomplish their vision.

You need to inject your work with the same creative motivation. Continue to work on your craft by building your own side project. Push boundaries in the code you use for your job and the code you use for yourself.

Just as painting is simultaneously work and play, so is coding. You can have a lot of fun exploring your craft. Build an app that does something no app has ever done before. Spend your weekends creating new games to play with your friends. Build a robot to clean your house, so you can put your feet up when you get home. Develop your own smart technology, so you have your own personal Alexa.

Software development offers you an unlimited canvas to work with. Your job is to keep pushing yourself to see the potential within your skills.

CONNECT THROUGH SOCIAL MEDIA

As I discussed in chapter 5, having a community around you is very important when trying to maintain motivation. As a software developer, you may or may not work with a team that keeps you engaged with your craft. One way to create that community outside the office is social media.

If you don't already have LinkedIn, Twitter, and Reddit accounts, set them up now. You can start to draw inspiration and get your questions answered even before you start learning to code. While you don't want to let coders on Twitter guide your career choices before you are in the industry, those same people can turn you on to some of the coolest, latest advances in coding once you've established yourself as a peer. Then you can contribute back to the community, helping out the next generation of coders by providing them with guidance on how to enter the industry.

This communication is mutually beneficial to everyone

involved, and it will help you maintain your motivation to keep learning and trying new things.

Social media isn't just a useful tool for socializing and motivation, either. Being a part of the online coder community can have a significant impact on your future career. Some of your fellow coders on LinkedIn and Reddit also run companies that may be looking for someone just like you. When they see that you engage well with others in the community and answer questions expertly, they'll know they're dealing with a rock star. And they may well want you working as part of their team.

A CAREER BEYOND CODING

Maintaining and sharing your craft can lead to some unexpected opportunities. Those positions may be at unique businesses or in incredible locations, or they may be in areas outside of software development entirely.

Not everyone who gets into coding feels passionate about the work. However, once you've proven you have the skills, new opportunities can arise that allow you to move away from the computer screen.

You may move into development management, taking a leadership role over a team of coders. You may move into software design, transitioning into a role where you come up with the concepts and get others to do the actual coding.

There are high-level positions in sales, human resources, and management for those who can demonstrate a deep understanding of coding. In fact, following this path can lead all the way to the top of major organizations. The current CEO of Microsoft, Satya Nadella, worked his way up from a software development position.

Within every tech company, there are leaders who started out just where you are now, with a vague understanding of the technology, a desire to learn, and an eagerness to achieve more in life. With those basic components, they moved up the ladder at their chosen company until they were in the boardroom with the likes of Mark Zuckerberg and Satya Nadella.

THERE'S ALWAYS ROOM TO TAKE CHANCES

You can climb the career ladder with software development, or you throw the ladder out and jump straight to the top of your own organization. If you're spending your weekends developing an app that no one has ever thought of, you may be creating the next Facebook.

This happens all the time. Dong Nguyen was just another software development student in Vietnam when he decided he wanted to make a simpler, Mario-inspired version of Angry Birds. He called it Flappy Bird, and it made him millions.

Minecraft came about in a similar way. A guy named Markus Persson made it for his company Mojang. After making hundreds of millions in profits, he sold Minecraft to Microsoft, where it remains one of the world's most popular games.

This doesn't just happen with video games. Uber was founded by Garrett Camp after spending $800 on a private car when he couldn't get a cab on New Year's Eve in Paris. He wanted to simplify and reduce the cost of getting around, so he created an app to do just that. Thus, a multibillion-dollar transportation company was born.

Almost every development in the tech world starts with someone just like you, sitting around, looking for a cool way to use their coding skills.

THIS IS JUST THE BEGINNING

These days, with a computer or phone screen never more than a few feet away, it can feel like the world was always this way. The fact is, we're only a couple decades into the computer age and the rise of the software developer. Most of the big tech companies we know by name today didn't even exist in the early 2000s.

In that short time, rock star coders around the world have set humanity on a new course, and because of those advances, software development isn't slowing down—it's speeding up.

In the years to come, every aspect of the world economy will involve online communication, automation, and AI. Because of the exponential increase in demand for efficiency from companies and immediacy from customers, there will continue to be an exponential need for software developers.

The code for all these ambitious projects won't write itself. People like you are going to have to build and maintain these technological solutions. If you think the world runs on code now, wait another decade. We're just getting started.

Of course, as technology evolves, software development will evolve with it. In thirty years' time, coders may not be using the .NET stack. They may code their new programs using new languages. But they will still be coding, and you can be one of them. So long as you commit to keeping up with the industry, you will change along with the technology, and your opportunities will only continue to grow.

That's why I'm confident that the next tech billionaire is reading this book. The power of these tools is growing, and the potential is rising. One of you will use those tools to grasp that potential and change the world.

Whether you're looking to be the next Zuckerberg or you just want a job you can trust to still exist by the time you retire, this is the only profession that can accommodate both dreams. Whatever you want from your career and wherever you want it to take you, software development can ensure you have an incredible future ahead of you.

All you have to do is to stick with it.

CONCLUSION

People say meritocracy in America is broken. That may be true in many industries, but there's one part of the economy where it's not: software development. If you follow the steps in this book, you really can rise as high as your ambitions and talents can take you.

Any clever, creative, dedicated individual can tap into this roadmap to reach the fulfilling, lucrative jobs we've discussed in this book. These jobs represent more than just a great income and the chance to live and work where you want, how you want, and for whom you want. They provide something missing in much of the modern economy: a real, profound sense of fulfillment.

One of my former students, Brent, developed an app for a real estate company. The app texted realtors when their appointments showed up. It was a relatively simple program, but you've never seen anyone as excited as Brent when he demonstrated it to me. He knew that as soon as his app went online, thousands of people were going to be using *his* creation every day.

I remember that same feeling when, years ago, I walked into a bank to see someone using TrueChecks®, the fraud software I wrote to check whether a transaction was fraudulent or not. I had to suppress a childish desire to walk up to the bank teller and say, "You know, I wrote that."

We usually associate that kind of creative fulfillment with those lucky few who become successful in the arts. We can imagine how we might feel if someone was reading our book in a café or listening to our music on the subway. We're taught that such satisfaction is for those rare, brilliant, popular artists, and not for us. We're told to put aside ideas of satisfaction, and to slog along just to get by.

You don't have to live like that. With software development, you can enjoy a sense of creative fulfillment, and you can say goodbye to the slog and the dull daily grind.

I hope you grasp this opportunity for yourself. I hope you go on the journey we've covered in this book and learn how to code. I hope you commit to this career and enter the one sector of the economy that can lead to your dreams.

If you're looking for the most efficient path to that success, visit us at CoderFoundry.com. There, you can take our Software Developer Career Quiz to find out how suited you are for this career. Then, you can sign up for our boot camp, where we have the tools to set you on this path faster than anyone else.

If Coder Foundry isn't right for you, find another boot camp, another mentor, and another path for yourself. You can always use our site, FiveStepsToASoftwareJob.com, to review what a good portfolio looks like.

No matter how you do it, remember that those telling you there's no hope for your future are wrong. You can follow the steps in this book to learn how to code, get that first job, and achieve your dreams. It's as simple as that.

ABOUT THE AUTHOR

BOBBY DAVIS JR. is a tech entrepreneur with over twenty years of experience building successful software companies. Bobby founded his first company, the custom software consultancy Core Techs, in 2002 with just $500 in his account. He has since grown it into a multimillion-dollar business. His second effort, Advanced Fraud Solutions, now runs in almost 1,000 financial institutions across forty-eight states. *Inc.* has labeled it one of the fastest-growing private businesses in the country four years running. Bobby also runs the Coder Foundry boot camp, where he has successfully placed hundreds of his software development students in high-paying jobs across the industry.

Printed in Great Britain
by Amazon

48763214R00106